SOTHIS PRESS

Sothis Press aims to completely reimagine the public domain book with an emphasis on universal access to captivatingly designed print editions. We believe that technology will enable every single book ever published to be perpetually accessible in beautiful print form. Sothis is beauty in ubiquity.

www.sothis.press
1 Waterhouse Square, London, ECIN 2ST

Tools, Chucks and Fixtures

A Comprehensive and Detailed Treatise Covering the Design and Use of Cutting Tools and Holding Devices Employed in Turning and Boring Operations in Modern Manufacturing Plants for Obtaining Accuracy and Increasing Production

Albert Atkins Dowd

SOTHIS PRESS

ISBN 978-1-0368-0980-5

This edition is published by
Sothis Press © 2023. The company
is registered in England and Wales.
The Company Number is 14900389.

TOOLS, CHUCKS

AND

FIXTURES

A COMPREHENSIVE AND DETAILED TREATISE COVERING THE DESIGN AND USE OF CUTTING TOOLS AND HOLDING DEVICES EMPLOYED IN TURNING AND BORING OPERATIONS IN MODERN MANUFACTURING PLANTS FOR OBTAINING ACCURACY AND INCREASING PRODUCTION

By ALBERT A. DOWD
PRODUCTION ENGINEER

FIRST EDITION
FIFTH PRINTING

NEW YORK
THE INDUSTRIAL PRESS
LONDON: THE MACHINERY PUBLISHING CO., LTD.
1920

Composition and Electrotyping by
F. H. Gilson Company, Boston, U.S.A.

PREFACE

THE developments in the design of machine tools during the past ten or fifteen years have brought these machines to a high degree of perfection. Many are provided with features which make it possible to obtain results, both as regards quantity and quality of output, which were practically unheard of a decade ago. It is obvious, however, that no matter how accurate and how well adapted to rapid production a machine may be, comparatively little is gained if the methods of holding the work and presenting the tools to it are not equally well thought out. As a matter of fact, these points are neglected in a great many machine shops. In a few instances we find planning departments and efficient tool-designing departments where the methods and appliances to be used in manufacturing are carefully considered. In the majority of shops, however, the workmen, or at best the foremen, are left to devise for themselves the methods by which the work is to be held and machined.

In order to aid these men, as well as the great number of tool-designers regularly employed on work of this character, the articles on tools and holding devices for engine and turret lathes, and horizontal and vertical boring machines, which have been published in MACHINERY by Mr. Albert A. Dowd, have been collected and are now published in this volume. These articles have been recognized as constituting the most fundamental and complete descriptions of tools and devices for machining operations ever published. In addition to the chapters applying specifically to turning and boring operations, considerable attention has been paid to the design of tools and holding devices, with reference to the cost of keeping them in repair — considerations which are of prime importance in economical shop management.

A very important feature of this book is that the arbors, chucks, fixtures and other holding devices shown and described are examples of designs actually employed in practice, and successfully applied to productive work in a great number of manufacturing establishments in this country. There is nothing of the theorist in Mr. Dowd's methods of describing the tools, chucks and fixtures dealt with. The descriptions are all plain matter-of-fact statements, and will, therefore, appeal with especial force to the great body of men of practical training who are seeking specific and up-to-date information.

A few of the chapters have already been published in MACHINERY'S well-known 25-cent Reference Books, and the present volume has been brought out in reply to a demand for a more comprehensive and detailed treatment than would be possible in these smaller books. It is the belief of the author and publishers that the work fills a well-defined gap in existing mechanical literature, and that the direct and specific method by which the information contained is given will meet with the approval of the trade for which this book has been prepared.

MACHINERY

New York, *January*, 1915

CHAPTER I

ADJUSTABLE AND MULTI-CUTTING TURNING TOOLS

CHAPTER II

DESIGN OF BORING TOOLS

CHAPTER III

RECESSING TOOLS

CHAPTER IV

FLOATING REAMER HOLDERS

CHAPTER V

ARBORS FOR TURNING, BORING AND GRINDING

Chapter VI

HOLDING DEVICES FOR LATHE AND BORING MILL WORK

Chapter VII

METHODS OF MACHINING THIN AND IRREGULAR WORK

CHAPTER VIII

TAPER BORING AND TURNING ATTACHMENTS

CHAPTER IX

MACHINING CONVEX AND CONCAVE SURFACES

CHAPTER X

METHODS FOR MACHINING ECCENTRIC WORK

CHAPTER XI

COUNTERBALANCED AND INDEXING FIXTURES

CHAPTER XII

INFLUENCE OF CHIPS ON THE DESIGN OF TOOLS AND FIXTURES

CHAPTER XIII

PROVIDING FOR UPKEEP IN THE DESIGN OF CUTTING TOOLS

TOOLS, CHUCKS AND FIXTURES

CHAPTER I

ADJUSTABLE AND MULTI-CUTTING TURNING TOOLS

Standard and Special Tools. — The cost of tool equipment for the manufacture of interchangeable work is an item which should be proportionate to the number of pieces to be machined. The saving in time which can be made by the use of special tools should also be carefully considered, as there are many cases where special equipments are designed for work which could be handled to advantage by the judicious use of standard tools. In order to obtain the greatest possible production from their machines, there have been instances where machine tool builders have sold tool equipments of expensive design, when a standard equipment would have done the work very nearly as well. Undoubtedly there was some gain in production, but it is doubtful whether the saving in time would pay for the special tools. The upkeep of special tools is also a factor which must be taken into consideration. It is interesting to note that the present aim of machine tool builders is to so design standard tool equipments that they can be adapted readily to a great variety of working conditions. A great deal of time is spent by manufacturers in devising and experimenting with various tools in order to perfect them to such an extent that they will conform to these conditions.

The rapid growth of the automobile industry in the past ten years is largely responsible for the broader development of our machine tools. The enormous quantities of interchangeable parts which are required in this industry and the manufacturers' desire for increased production have brought into existence a great variety of multi-cutting tools. Tools of this kind may be designed for a variety of uses, and tool-holders capable of containing several tools can also be designed for handling a considerable range of work.

Tools for Turret Lathes and Boring Mills. — Adjustable tools and those having cutters for turning several diameters are sometimes combined with boring-bars, drills, or cutter heads, these being applied to some one of the various types of turret lathes. They are also occasionally designed for use on a vertical boring mill.

When used on the turret lathe, the cut-off slide is frequently equipped with a gang of tools so that the operations of turning, boring and facing can be carried on at the same time. Quite frequently the tools are so arranged that from nine to twelve are working at the same time, with the result that there is a considerable gain in production. There are a great many varieties of so-called " box-tools " on the market, and these are principally used for bar work on turret lathes or screw machines having a collet mechanism. Tools of this type are usually a part of the standard equipment furnished with screw machines adapted to bar work, and they will not be discussed in this book.

Design of Multi-cutting Turning Tools. — The design of multi-cutting turning tools for castings and forgings which have several diameters to be machined is a subject well worth considering, for it is safe to say that nearly any manufacturer who uses horizontal or vertical turret lathes can greatly increase the productive efficiency of his machines by the judicious use of multi-cutting tools. The several designs of turning tools illustrated in this chapter have been built for various purposes, and a careful study of the types shown may be of assistance in suggesting methods which can be used to perform some piece of work requiring tools of a similar kind. Some of the important points in the design of tools of this nature are as follows:

1. Rigidity: Avoid overhang as much as possible unless some sort of outboard support can be used. Pilot the tool if practicable.

2. Arrangement of tools: The tools should be perpendicular to the plane in which the turret rotates when indexing, because variations in diameters are less likely to take place when tools are arranged in this way. Unequal indexing of the turret has very little effect on the radial position of the tools under these conditions, so that the sizes can be obtained much more exact than if the tools are placed in a plane parallel to the turret rotation. Use standard rectangular stock for the cutting tools so that the upkeep will be inexpensive and reforging be avoided.

3. Try to make the block containing the tools removable, so that it can be replaced easily by another block with tools arranged differently to handle other work.

4. Make the tool-block adjustable if possible.

5. Back up the tools with adjusting screws.

6. Make provisions so that cutting lubricant can be directed on the faces of the tools when forgings or steel castings are to be machined.

7. Arrange the tool-block in such a way that the thrust of the cut does not come against it; it is much better to have the thrust come on the body of the tool.

Multi-turning Tool for Electric Motor Shafts. — One example shown in Fig. 1 is given of a multi-turning tool for bar work. This tool was designed for use on the electric motor shaft shown at *A* in the illustration. The work was handled in short lengths although the stock is a regular commercial product. Roughing and finishing operations were performed with the same type of tool. The work was held in collet jaws. Something like twenty varieties of shafts having different diameters and shoulder lengths were handled by these tools.

A Pratt & Whitney turret lathe with collet mechanism was used for this work. As this type of machine has a turret with dovetail faces, the body of the tool *G* was arranged to fit the dovetail and the gib *S* held it in place. The cut-off slide was planed off at the center and the hardened steel block F was secured to it. It will be noted that this block acts as a support for the tool, and the tongue assists in preventing lateral movement. The cast-iron block *Q* is fastened to the body of the tool and it is dovetailed at *U* to receive the tool-slide *P*. This is of steel and it is T-slotted so that standard toolposts *O* can be used. It will be seen that the tools *C*, *D* and *E* are held in such a way that they can be adjusted readily both for different diameters and for shoulders of varying lengths. The slide is screw controlled and is operated by the handle *T*. A

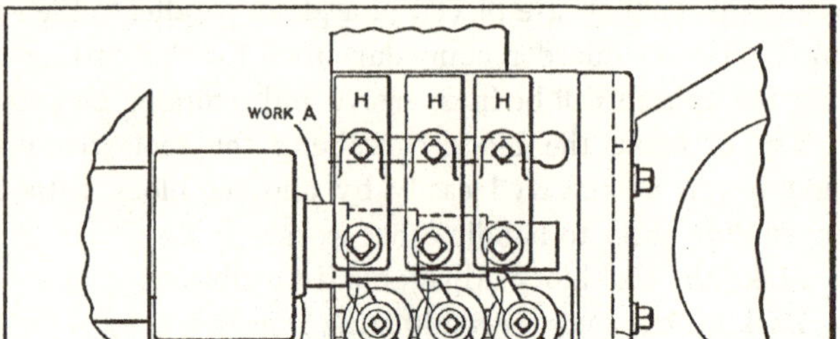

Fig. 1. Special Multi-turning Tool for Bar Work

positive adjustable stop is provided by the check-nuts *V*. The back-rests *J* are of tool steel and are contained in the brackets *H*. The screws *K*, *L* and *N* are used for binding and adjusting, while those at *M* pass down through slots in the body of the tool and permit adjustment of the back-rests in a longitudinal direction.

Points worthy of notice in this tool are the method of supporting the body by means of the block on the cut-off slide, the flexibility of the tool adjustments and the ease with which any tool may be replaced if broken or used up. The tools are of standard section and therefore require no machining except cutting off and grinding.

Multiple Tool-holder for the Turntable Type of Lathe. — The bronze nut shown at *A* in Fig. 2 is an example of a piece

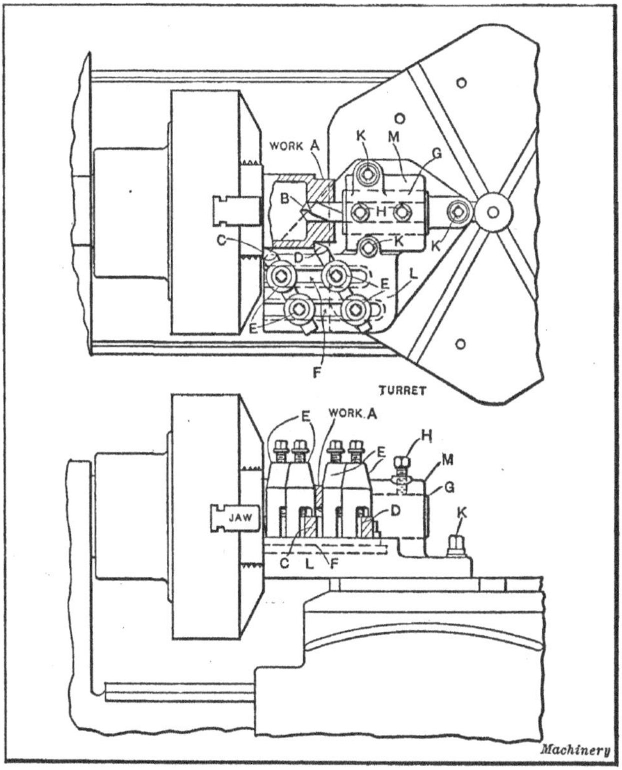

Fig. 2. Multiple Tool-holder for the Turntable Type of Lathe

of work which is to be drilled and turned on two diameters at the same time. There were six pieces of this kind varying slightly in size, which were to be machined in lots of twenty-five. Two tool-holders were used to do the work, one tool being arranged as shown in the illustration, while the other contained a boring tool in place of the drill *B*. The holder *L* was made of cast steel and was T-slotted in two places at F, so that tool-holders *E* of the standard type could be used. These carry the tools *C* and *D*, and attention is called to the way in which two posts are used for each tool to insure the maximum rigidity. The body of the holder is fastened to the turret face by the

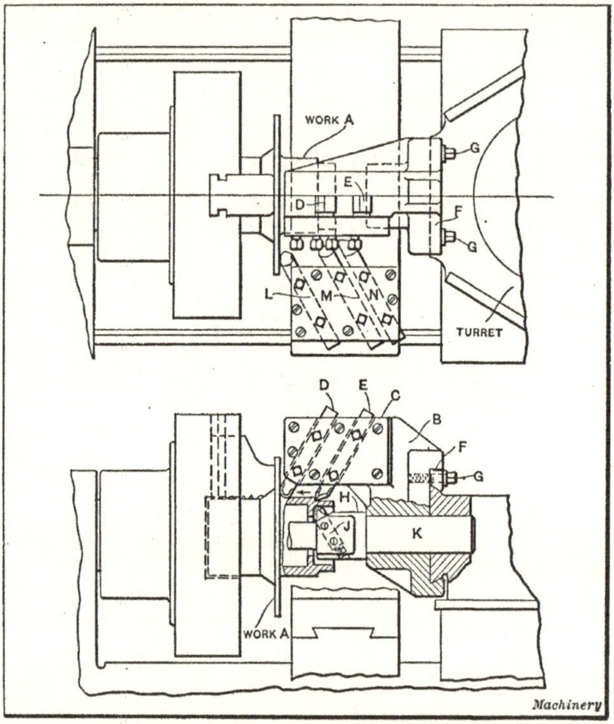

Fig. 3. Multiple Tool-holder for Machining Automobile Hubs

screws *K*, and is tongued on its under side to fit the slot. A semi-cylindrical boss *M* contains the split bushing *G* which holds the drill *B*. Two screws *H* are employed to clamp the bushing. This holder is simple, easily made and quite adaptable for work within its capacity. There are likely to be slight variations in the diameters turned due to imperfect indexing of the turret, but for general commercial work these usually are not great enough to cause any serious trouble.

Multiple Tool-holder for an Automobile Hub. — The piece of work shown at *A* in Fig. 3 is an automobile hub, and the tool-holder is arranged so that the operations of turning and boring can be carried on simultaneously with the facing. The tools *L*, *M* and *N* are secured in a special block on the cut-off slide. The tool-holder *B* is of cast iron and well ribbed; it fits the dovetailed face of the turret, being secured in position

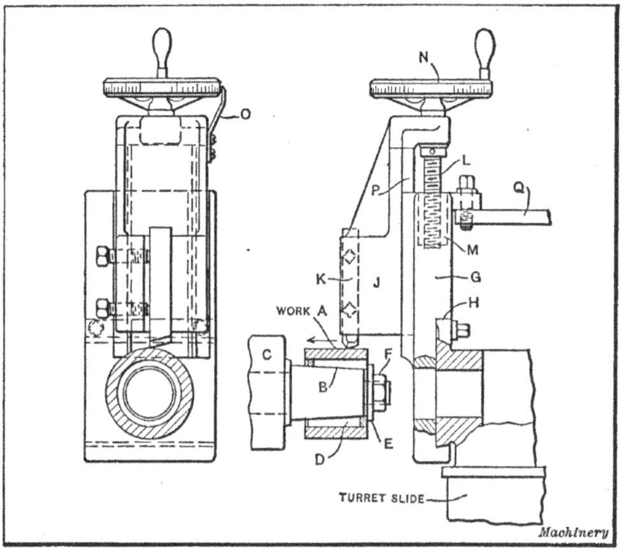

Fig. 4. Adjustable Turning Tool for Finishing the Outside of Short Bushings

by the screws *G* and the gib F. The turning tools *D* and *E* are mounted vertically, and the steel cap-plate *C* contains the necessary holding screws. The boring-bar *H* is piloted in a chuck bushing at its forward end and contains the tool *J*, which stands in a vertical plane like the turning tools. The shank of the bar *K* is secured by the turret binding screw and an additional set-screw in the holder itself. A tool of this type will produce more accurate work than the type shown in the preceding illustration, on account of the position of the cutting tools with reference to the turret indexing. A feature of some importance is the piloting of the boring-bar, as this assists in the prevention of vibration. Care should be taken in the design of tool-holders of this type, that the overhang from the turret face is not too great, for if this is excessive, a certain amount of chatter is inevitable.

Adjustable Turning Tool for Short Bushings. — A number of short bushings, such as that shown at *A* in Fig. 4, were to

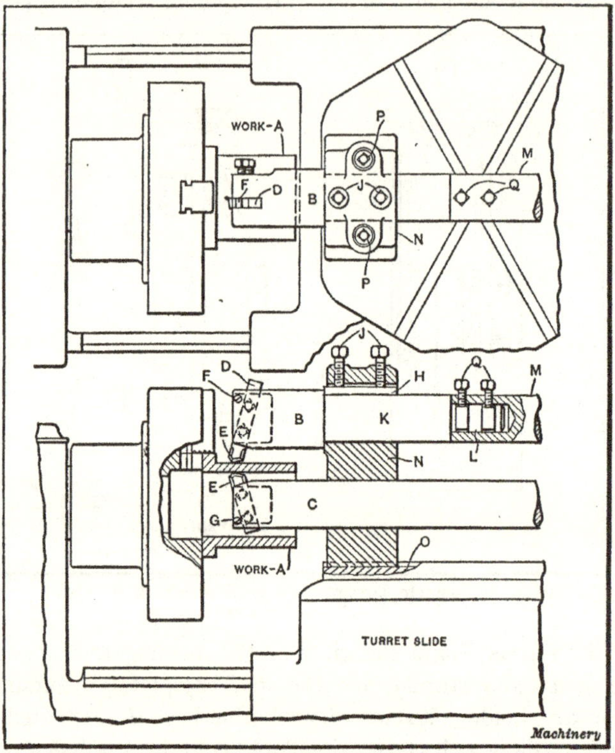

Fig. 5. Boring-bars for Turning Concentric Packing Rings

be refinished on the outside. The bushings were of various diameters ranging from $2\frac{1}{4}$ to 4 inches, while the lengths varied from $1\frac{1}{2}$ to 3 inches. The pieces were held on arbors *B*, in collet jaws *C*. A split sleeve *D* was expanded inside the work by the action of collar *E* and nut *F*. The body of the tool *G* was secured to the dovetailed face of the turret by gib *H*. The tool-slide *J* is a steel casting dovetailed at *P* and fitted with an adjustable taper gib to take up the wear. The cutting tool *K* is placed in a slot in the slide and is secured by the screws shown. The screw *L* is journaled in a lug at the upper end of the slide and enters a steel nut *M* in the body of the tool. Radial adjustment is obtained through this screw by means of the hand-wheel *N*. The rim of the wheel is graduated and the pointer *O* permits accurate readings to be made. This tool is very good indeed for light work, and accurate results can be obtained by its use. When two tools are used, a tie-rod *Q* assists in making the combination more rigid.

Holder for **Adjustable Boring- and Turning-bars.** — The work shown at *A* in Fig. 5 is a cast-iron pot from which concentric packing rings are to be cut, and the boring and turning are done at the same time. Two cast-iron holders *N* are tongued at *O* and secured to opposite sides of the turret by the screws *P*. The turning- and boring-bars *B* and *C* pass through the holders and extend entirely across the turret. The boring-bar *C* is of the same diameter along its entire length, and it is secured in the holders by shoe binders similar to that shown at *H* but located in the sides of the holders. The boring tools *E* are of rectangular section and secured by set-screws in the slots at the ends of the bar. The screws F and *G* help to stiffen the ends of the bars where they are slotted. The upper or turning-bar is made in two sections *K* and *M* so that the tools may be swung radially to bring them into their proper position when the turret is set off center for turning larger diameters. The end of one bar is turned down at *L* to fit the hole in the other bar and the screws *Q* make a firm joint. A set of bars and holders of this kind is a very useful adjunct to a turret lathe equipment, and it may be adapted to a variety of uses. The double tie feature across the turret gives exceptional rigidity.

Piloted Multiple Turning Tool for **Steel Gear Blanks.** — The automobile jack-shaft gear blank shown at *A* in Fig. 6 is of alloy steel and is held in special chuck jaws so that it can be drilled, turned and faced simultaneously. A special tool-block on the cut-off slide performs the latter operation, while the turning and drilling tools are carried by the turret. The body of the turning tool *V* is made of cast iron and is fastened to the dovetail turret face by the gib shown at *U*. The tool-block *K* is of steel and is slotted to receive tools *M* and *N*. An oil groove is cut at *L* along the top of the block and it is supplied with oil from the special piping system shown. The

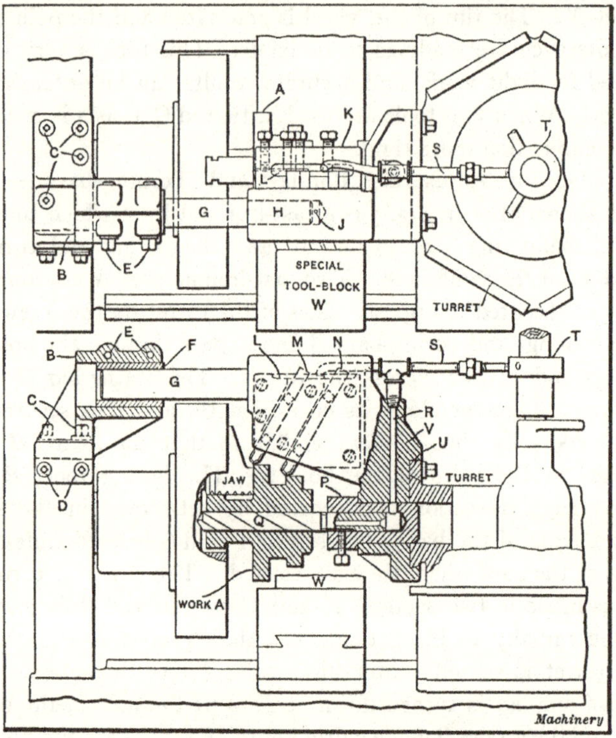

Fig. 6. Piloted Multiple Turning Tool for Gear Blanks

pipe *S* leads to the distributing collar *T* which, in turn, is connected with the cutting lubricant piping system on the machine. The slots in the tool-block are of sufficient width to permit an ample supply of fluid to run down and reach the cutting ends of the tools, thus assisting greatly in prolonging the life of the tools and also allowing higher cutting speeds.

The oil drill *Q* is held in a steel supporting bushing *P* which fits the body of the tool-holder. It is supplied with lubricant through the hole *R* which is connected to the piping system. The steel pilot *G* is shouldered at *H* and is forced into the body of the holder. The small hole *J* is put in so that air pressure will not be generated when the pilot is pressed into place, as this would tend to deceive the fitter by making him think he had a good fit when, in reality, it was compressed air that made the pilot hard to force in. The author has seen pilots fitted so that they were apparently all right at the time when the work was done, and yet when the time came for the tools to be used, it was found that they were loose enough to cause trouble. The air hole will prevent trouble of this kind.

A special bracket is shown at *B* and it is screwed to the spindle cap by the screws *C* and *D*. The bronze bushing F receives the end of pilot *G*, and it is clamped by the binding screws *E*. This method of supporting a turning tool is very successful and assists greatly in permitting heavy cutting without chatter. Another feature of this tool is the manner in which oil is conveyed to the cutting tools. Attention is also called to the position of the tool-block, this being at the rear of the body so that the thrust of the cut is brought directly against the heavier part of the casting. The method of mounting the tools is also a little out of the ordinary, in that the block and tools form a unit which can readily be removed, permitting the substitution of another block with tools arranged differently, to handle other work requiring different spacing. Two turning tools on opposite sides of the turret were used for this particular piece, one being used for roughing and the other for finishing.

Double-end Piloted Turning Tool for Pistons. — The piston *A* shown in Fig. 7 is held by the inside on a special expanding pin chuck *B*. The arrangement of tools is that recommended by the Pratt & Whitney Co., for turning automobile pistons on their horizontal turret lathe. The turning tool-holder *C* is of cast iron and is double ended, reaching entirely across the turret, and the two ends are exactly the same. The body or arm is of U-section and it is cored out at the center so that the turret binder-lever can pass through it, as shown at *G*. A careful fit is made on the dovetail at *D*, and two special bolts F pass down through the body of the tool and pull up the gib *E* against the lower dovetail, thus clamping the tool securely. The tool *J* is backed up by the collar-head screw *H* and is

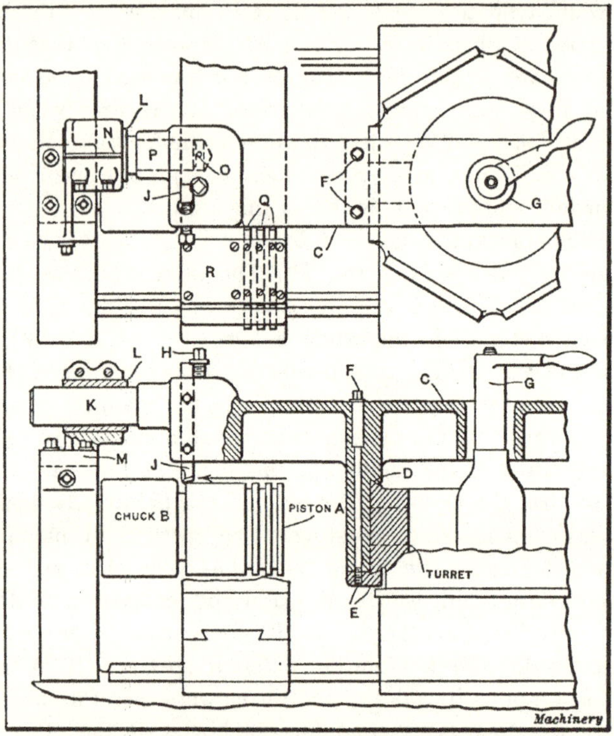

Fig. 7. Double-end Piloted Turning Tool for Pistons

secured by means of two screws. The steel pilot *K* fits the bushing *L* in the spindle cap bracket *M*, as in a previous instance. It will be noted that an air hole *O* is provided where the shank *P* enters the end of the tool body. In connection with the turning tools a special block on the cut-off slide is used to cut the ring groove in the piston. This block and tools are shown at *R* and *Q* in the upper view. This style of equipment is very well known and has given universal satisfaction.

Piston Turning Tool having Adjustable Tool-block. — A development of the preceding tool is shown in Fig. 8. It will be noted that although the general construction is about the same, in this instance the tool-block is made separate so that other blocks may be substituted having more than one tool.

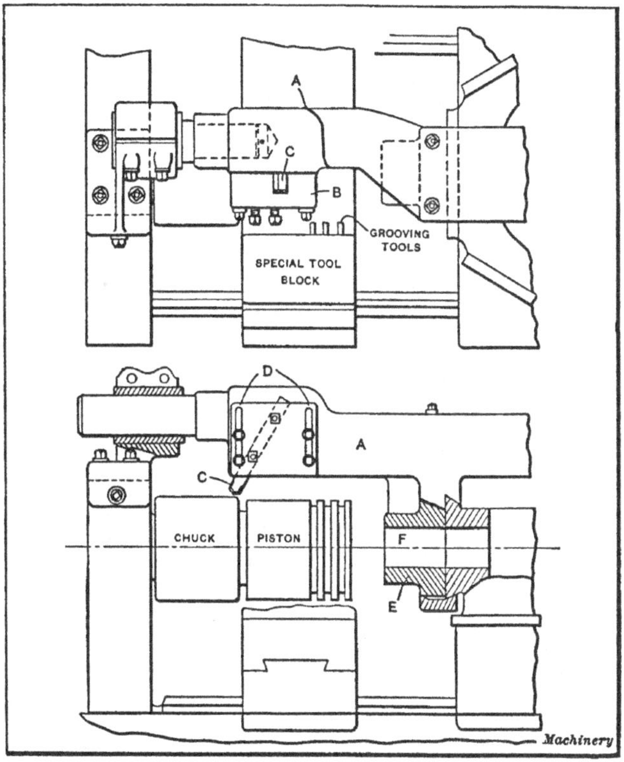

Fig. 8. Piston Turning Tool having Adjustable Tool-block

Considerable adjustment is also permissible by means of slots shown at *D*. It is obvious that this method of construction requires the tool-block *B* to be of steel and somewhat heavy, so that it will properly resist the thrust of the cut. The screws which hold the block in position must also be of ample size. As this particular tool was designed for use in turning and boring

ring pots, in addition to piston work, the boss *E* was supplied and bored at *F* to receive a boring-bar.

Special Multiple Turning Tool with Roller Back-rest. — In a great many instances the design of an automobile piston is such that it is permissible to center the solid end, and this

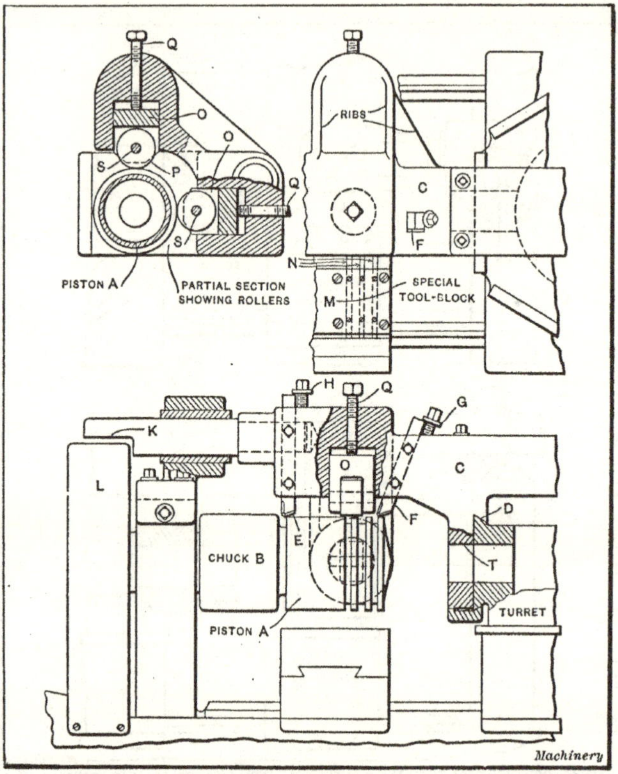

Fig. 9. Multiple Turning Tool equipped with Pilot and Roller Back-rest

gives a chance to support the end by a conical rest. While the ring grooves are being cut some support is essential, and in the case of the piston shown in Fig. 9 the use of roller supports in place of a center rest was found necessary for the reason that centering was not permitted. The piston *A* is held on the special chuck *B* and the two tools *E* and *F* are held in a double-end tool-holder. Adjustment is obtained by means of the collar-head screws *H* and *G*. The turning tool body *C* fits the turret dovetail at *D* and it is clamped, as previously stated. The end of the pilot *K* is cut away on its under side in order to clear the gear guard *L*. The steel supports *O* are backed

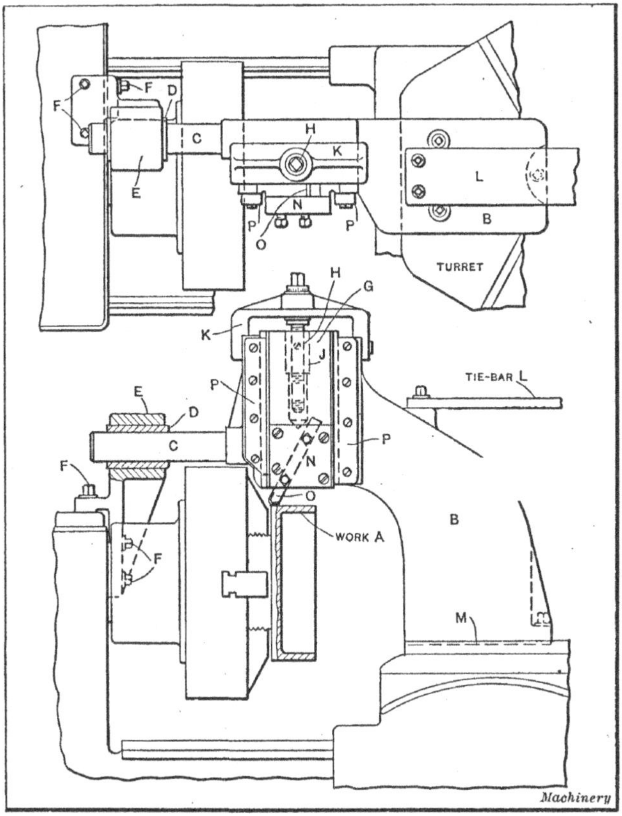

Fig. 10. Adjustable Piloted Turning Tool for Large Work

up by the screws *Q*, which are also used for adjusting purposes. The hardened and ground steel rollers *P* are hung on the pins *S*. (See detail view.) A special tool-block *M* contains the grooving tools *N*. This equipment also was very successful.

Adjustable Piloted Turning Tool for Large Diameters. — A somewhat different type of tool is shown in Fig. 10, this being adjustable for various diameters from the 12-inch casting *A* down to a diameter of 6 inches, or a trifle under that size if necessary. This tool was rather heavy and cumbersome and not entirely successful on heavy cuts. On the lighter variety of work, however, it proved satisfactory and adaptable. Two tools were used on opposite sides of the turret; the flat steel tie-bar *L* helped to prevent sagging.

The body of the holder *B* is of cast iron cored out so that the walls are ½-inch section, and it is tongued along its lower face

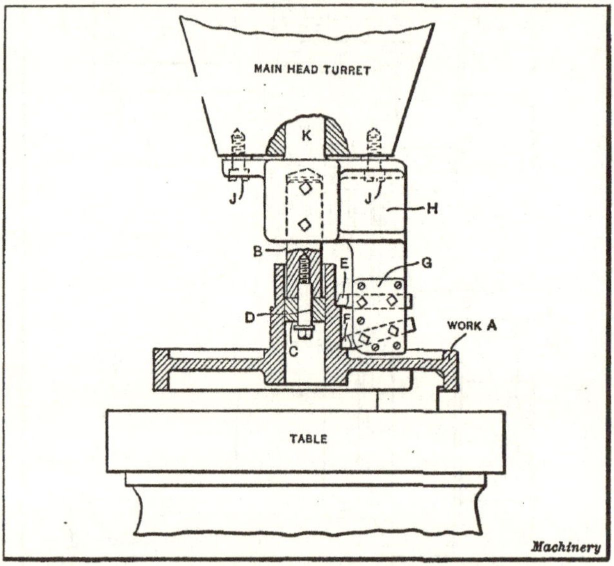

Fig. 11. Multiple Turning Tool for Vertical Turret Lathe

to fit the turret at *M*. The forward end holds the steel pilot *C*, which is supported and guided by the bushing *D*. The bracket *E* is fastened to the head of the machine by the screws *F*, thus insuring a rigid support for the end of the pilot. The tool-slide *N* contains the tool *O* and it is securely gibbed by the two steel straps *P*. A taper gib (not shown) provides adjustment for wear on the sides. The bracket *K* is screwed to the top of the tool body and journals the operating screw *H*. A graduated collar permits accurate settings to be made without trouble. A bronze nut in the body of the slide at *J* receives the operating screw.

Multiple Turning Tool for the Vertical Turret Lathe. — The vertical turret lathe is less frequently supplied with multiple tools than the horizontal type of machine, for the reason that the regular equipment supplied by the manufacturers is adapted to a wide range of conditions without very much special tooling, and, in addition, the class of work for which this machine is more likely to be used is of such a nature that multiple turning

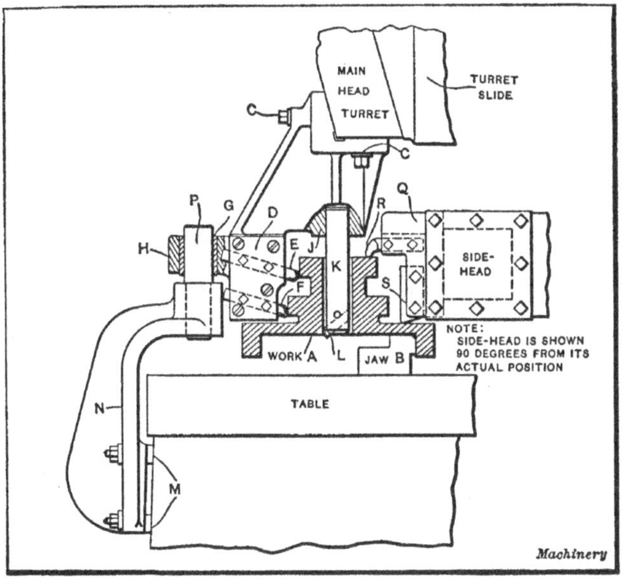

Fig. 12. Piloted Multiple Turning Tool for Triple Gear Blank

tools are less likely to be required. There are instances, however, when a considerable increase in production may be made by the use of the multiple type of tools. Take for example the special gear shown at *A* in Fig. 11. This piece of work is held by the inside of the rim in special jaws, and the tools in the side-head turret are used to face and turn the gear portion while the special multiple tools are at work on the hub. Before the operation illustrated takes place, the work has been bored, reamed and faced on the other side.

The body of the tool *H* is of cast iron and it is fastened to the turret face by the screws *J*, while the plug *K* centers it in the turret hole. The turning tools *E* and F are secured in the slots and a steel cover-plate *G* gives support for the set-screws which hold the tools in place. A steel shank *B* has a revolving roll *C* fastened to its lower end by the shouldered screw *D*; this roll acts as a pilot in the finished hole. The construction of this device is simple and the results obtained by its use are excellent.

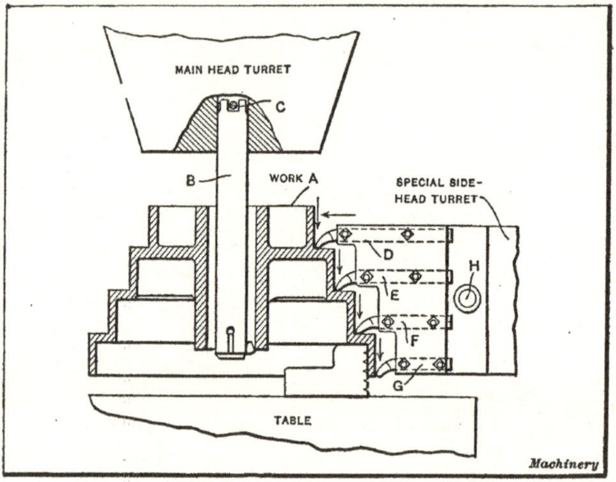

Fig. 13. Multiple Toolpost Turret for the Side-head

Piloted Multiple Turning Tool for Triple Gear Blank.— A radical departure from regular methods is shown in Fig. 12. The work *A* for which the equipment was designed is a cast-iron triple gear blank. Attention is called to the fact that in this illustration the side-head is shown in a false position in order to show the cutting action more clearly. The body *H* of the multiple turning tool is fitted to the turret and held in position by the screws *C*. A steel bushing *G* acts as an outboard support for the tool, and it is a sliding fit on the pilot *P* which is shouldered in the supporting bracket *N*. This bracket is heavily ribbed and is fastened to the bed of the machine. The adjustable washers at *M* are used to align the bracket properly. A tool-block *D* contains the two turning tools *E* and *F*, and the boring-bar *K* is held in the hub *J*. The arrangement of the side-head, in this instance, is a little out of the ordinary. A special tool-block *Q* contains the two facing tools *R* and *S*, and these work simultaneously with the turning tools, thereby making production very rapid.

Multiple Toolpost Turret for the Side-head. — The cone pulley shown at *A* in Fig. 13 was machined in one setting. The

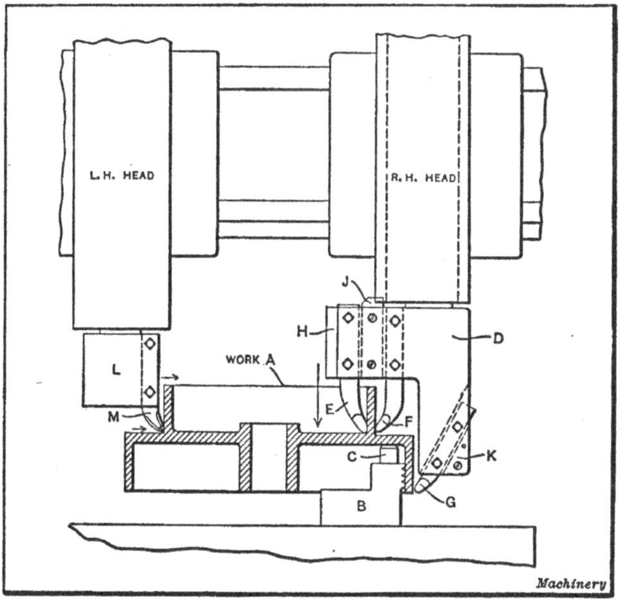

Fig. 14. Application of Multiple Turning Tool to Vertical Boring Mill

casting was held by the inside of the lower or largest step of the cone and a driver (not shown) was placed against the interior ribbing, as the jaws were not sufficient to hold the work securely against the cutting action of the four turning tools. A special side-head turret tool-holder was designed for this piece of work, and the facing and turning tools *D*, *E*, *F* and *G* were held in it as shown in the illustration. One set of tools was used for roughing and a duplicate set on the other side of the turret post was used for finishing. The entire group of tools pivoted on the stud *H*. While these cutters were operating on the outside of the pulley, the boring-bar *B* (held in the main-head turret and driven by the pin *C*) was slowly boring out the hole. A forming plate was used to give the desired crown to the steps. The production could have been improved if a special turning tool had been used in the main head for turning the four steps of the cone, and the side-head used for facing only. These operations could have taken place at the same time, and the speeds would have been more nearly correct. The boring could have been done at a higher rate of speed. However, the results obtained with the arrangement shown in the illustration were satisfactory.

Multiple Turning Tool for the Vertical Boring Mill. — The vertical boring mill is seldom equipped with multiple turning tools, but there are cases where production can be increased considerably by their use. One example only is given of the use of an equipment of this kind. Fig. 14 shows a large pulley at *A*, and this is held by the inside of the larger step in the special jaws *B*. The buttons *C* give a three-point support to the work. A special tool-holder *D* is slotted out to receive the tools *E*, F and *G* which are used for the turning and boring. The plate *H* is fastened over one end of the tool-holder in order to tie it together, and the filler-block *J* gives additional strength while its upper end engages the right-hand ram and acts as a driver. Another block *K* ties the lower end of the tool-holder together. The left-hand head contains the toolpost *L* which supports the tool *M*. This tool is used for facing while the other tools are turning.

Other instances of multiple turning might be given, and illustrations shown, but these would be much on the same order as those which have been mentioned and would be of no particular value as representative types. Tools have been selected for this chapter which seem to best illustrate the principles of design required in the various types.

CHAPTER II

DESIGN OF BORING TOOLS

Factors that Influence the Design of Boring Tools. — A boring tool or boring-bar is, in itself, a very simple tool and yet, in its various applications, it may require considerable forethought in order to obtain a tool which will be exactly the right one for the job. In order to properly design any kind of a cutting tool, an intimate knowledge of the actual working conditions which are met with in using the tool is a valuable asset. There are a number of factors which influence the design of boring tools and there are also many types of machinery to which boring tools may be applied. In some cases the bar revolves with the spindle of the machine, while in others it is held rigidly and the work revolves around it. These things affect the design and must be considered. The work naturally controls the size of the bar and also its shape, while the material which is to be cut makes a difference in the shape of the tool and determines the amount of "chip clearance" necessary.

The tools described and illustrated in this chapter must be considered as representative types of the many varieties to be met with in the general course of manufacturing. Points in design and construction will be noted and faulty tools will be discussed and criticised.

General Points in Boring Tool Design. — Some of the important points in the design and construction of tools for single-point boring are here given, and while some of these may seem obvious, they may be of assistance in calling attention to matters which would otherwise be overlooked.

1. Chip clearance must be very carefully looked after when the tool is to be used for cutting steel, as an accumulation of chips caused by insufficient clearance is very annoying to the operator and also injures the work by tearing or scratching it,

and finally ruins the bar itself unless it is hardened. The amount of clearance between the bar and the work should be as great as possible without sacrificing strength, and in this connection it should be noted that in addition to the necessary chip clearance at the point where the cutting action takes place, provision must also be made to get rid of the chips themselves. For this reason the portion of the bar beyond the cutting tool should be so proportioned that chips will not wedge. In cutting materials other than steel the clearance is not so important, as the chips are short and do not curl up or cling to the bar, so that they practically take care of themselves.

2. The method of holding or clamping the tool in position should be such that the thrust of the cut comes against the solid body of the bar and not against the set-screws or clamps. It is advisable to use square-head set-screws instead of the headless type whenever possible.

3. Boring-bars should be provided with some means of adjusting the tools for diameters, by the use of "backing-up" screws or wedges. The so-called "sledge hammer adjustment" type of bar should never be used when there is room enough to put in adjusting screws.

4. Boring-bar tools should be made as large as the diameter of the bar will permit without sacrificing strength, in order to assist in carrying away the heat generated by the cutting action, and to permit the use of heavier feeds without burning the tool. The rake of the tool should be such that it will turn the chips to the best advantage.

5. The bar should be so designed that micrometers can be used over the bar and tool in order that the operator may be able to set his diameters closely without resorting to the usual "cut-and-try" method used by our forefathers.

6. In the design of multiple boring-bars which are to be used to bore up to a shoulder, it is not good practice to set the tools in the bar at an angle. They should be located in a plane perpendicular to the axis of the bar. If set at an angle it will be found a very difficult matter to grind the tools so that diameters and shoulder distances will remain constant.

7. Bars designed for use on turret lathes should have the tools set in a plane perpendicular to the rotation of the turret. By this means variations in the indexing of the turret are minimized in their relation to the cutting tools, so that diameters can be held much closer to size than if the tools are arranged in a plane parallel to the turret rotation.

8. When the work is of such a nature that a cutting lubricant is required, provision should be made so that an ample supply of the fluid can be carried directly onto the face of the cutting tool. This result can be accomplished either by means of a hole in the bar with outlets at the proper places, or oil grooves covered with a strip of sheet brass. In either case a good connection must be made with the cutting lubricant system on the machine. This may be arranged by a distributing collar on the turret or by means of a special oiling device through the spindle.

Boring Tools for the Engine Lathe. — Boring tools which are designed for use in the engine lathe are generally of a very simple kind, adapted only to light cutting and seldom used for more than one or two pieces of work of the same size at the same time. Several varieties are to be found in the average toolroom, although forged tools will be noted in greater numbers than any of the others. Tools of this kind of almost every conceivable shape and size, from a small round "hook tool" for cutting an inside recess, to a large bar of tool steel bent over at the end for boring some long pieces of work, will be found in abundance. There are square bars and round bars with inserted tools, and, in addition to these, each toolmaker has a special boring tool of his own make which he uses for jig work. These special tools occasionally show considerable ingenuity in their construction, and are usually made in such a way that very fine adjustments can be attained.

The upper part of Fig. 1 shows a piece of work *A* held by the outside in chuck jaws, the machine on which the work is to be done being an engine lathe. A plain forged tool *B* is held in the toolpost *C* on the cross-slide of the lathe. This type of tool is the simplest of all tools used for boring and consists of a rectangular piece of tool steel of suitable size to fit the toolpost. The tool is drawn out and bent over at the cutting end by the blacksmith and is then ground to a cutting edge by the workman using it.

Hundreds of tools of this variety can be found in every machine shop and factory in this country. They are suitable only for light cutting and there is a tendency toward "chatter"

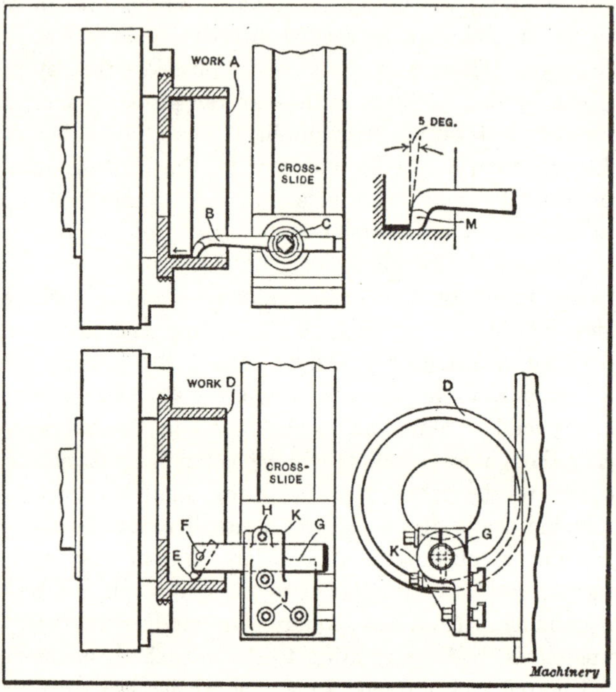

Fig. 1. (Upper View) Forged Type of Boring Tool; (Lower View) Boring Tool with Inserted Cutter

even when the cut is light; this is due partly to the shape of the cutting end and partly to the overhang of the entire tool. It will be found that less chatter will result if a slight land or flat is stoned on the tool immediately below the cutting edge. The tools should also be set slightly above the center. On casting where scale is encountered, there is a decided tendency for the tool to ride up on the scale and ruin very rapidly if it is ground as shown at *B*. The enlarged view *M* shows another method of grinding which is useful in cases of this sort. It will be noted that there is a slight back taper to the end of the tool and this assists in preventing any riding up on the scale, as its tendency is to make the cutting point draw in slightly and thus keep under the scale. Care must be taken not to make the angle too great — 5 degrees is ample, and much less than this can be used if desired.

The lower part of Fig. 1 shows the same piece of work *D* with another type of boring tool in action. A cast-iron body *K* is held down on the cross-slide of the lathe by means of the three bolts *J*. A steel bar *G* is longitudinally adjustable in the cylindrical portion of the holder and is clamped in position by means of the binder screw *H*. A round cutting tool *E* is held in place by the taper pin F, in a manner familiar to all. A holder of this type will be found a very useful adjunct to any tool-room, and is adaptable to a variety of conditions. A series of bushings can be made to take different diameters of round stock, and tools may be quickly made to suit almost any case. Obviously, adjustment for diameters is made by the cross-slide. Rigidity and adaptability are points in favor of this device.

Adjustable Boring Tools for Jig Work. — Fig. 2 represents two styles of adjustable boring tools used mostly for boring small shallow holes or jig bushings. These tools are capable of fine adjustments but are not suited for any kind of heavy cutting. The upper part of the illustration shows how tool *B* is used for boring a part of the bushing *A*, which is held in chuck jaws. The body of the tool-holder *H* is made of steel and is turned down and tapered at *J* to fit the tailstock spindle. The adjustable swivel *D* is pivoted on the shouldered screw *E*, and is adjusted by the two headless set-screws F and *G*. The tool *B* is of round section and fits the end of the swivel, where it is held in place by the two screws *C*. The end of the tool is bent over for the purpose of clearance. A tool of this kind is very convenient and is easily adjusted for diameters within its capacity. It is not adapted to deep holes, but can be made up in several sizes so that it will handle fairly large work.

The lower part of the illustration shows another tool of somewhat similar construction, which is designed for the same purpose as the other. The body is of steel and the shank *T* is turned taper to fit the tailstock spindle. The forward portion of the body *R* is cut out to receive the swivel *N*, which pivots

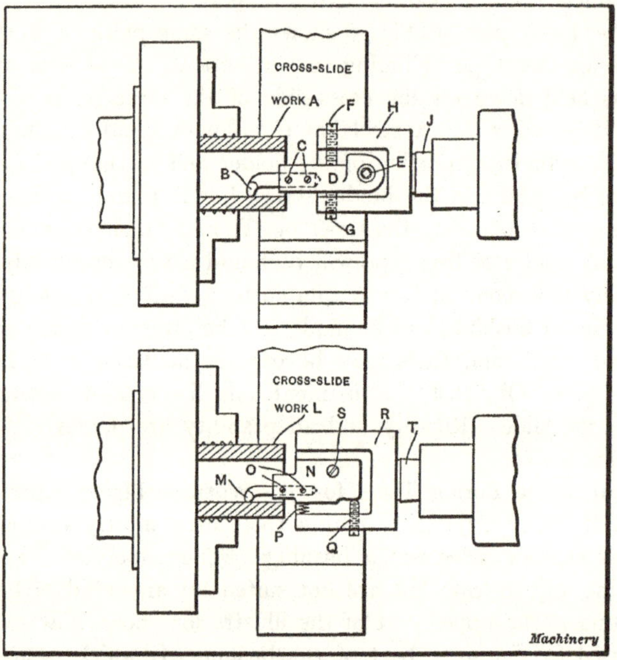

Fig. 2. Two Types of Adjustable Boring Tools for Tool-room Work

on the screw *S*. The tool *M* is of round section, bent over at the end, and it is held in place by the two screws *O*. One adjusting screw *Q* is all that is required in this tool, as the coil spring *P* takes up lost motion and prevents drawing in. This tool is not as rigid as the one previously referred to, but the spring makes adjusting much quicker, as only one screw is needed. A number of tools of this type, and of various sizes, were made for tool-room use in a large automobile factory and were used on the greater part of the jig work.

Boring Tools for the Horizontal Turret Lathe. — Boring tools which are required for use on the horizontal turret lathe are of many forms and their design is somewhat dependent on the type of machine to which they are to be attached. On machines having no transverse movement to the turret slide, the tools are nearly always designed for straight boring, while on the other types of machines, *i.e.*, those having transverse movement, the design is frequently made in such a way that the tools can also be used for facing operations. The form of the turret itself also influences the design to a certain extent, for it is evident that a flat turret would require a different type of tool-holder than one of the vertical face variety.

Single-point Starting Tool for Taper Holes. — The work *A* shown in Fig. 3 is a malleable iron automobile hub with a cored taper hole which runs out of truth very badly. Therefore it was necessary to design a starting tool of the single-point variety in order to generate a true running hole, so that the subsequent tool would start properly without being influenced by the wobble of the core. This tool and tool-holder are very simple, the tool itself being a piece of round high-speed steel bent over on the end and ground to cut a diameter a trifle smaller than the large end of the tapered hole. The holder *E* is a piece of machine steel of cylindrical shape, which is ground on the outside to fit the turret hole and on the inside to fit the shank of the tool *B*. Two set-screws *D* are used to hold the tool in position. It will be noted that the end of the cutting point is ground very nearly square so that it will not ride up on the scale. The tool is not made for continuous boring but is merely used to generate a true hole for a short distance into the cored portion of the hub.

Boring and Facing Tool for a Flat Turret. — An example of a boring tool which is also used for facing out a pocket on a turret lathe having a flat turret is shown in the lower part of Fig. 3. This tool is of the "shovel nose" type and its cutting action is rather hard on account of the bluntness of the nose and the amount of stock which it removes. The work *F* is a machine steel forging and the shoulder is not recessed at all in the blank. The tool *G* is of rectangular section and it is forged and ground on the cutting end to the shape shown. The tool-holder *H* is supported by the steel bracket *J* which is fastened down on the turret face by screws *L*. The slots *K* are T-shaped and permit various settings and combinations of tools to be made.

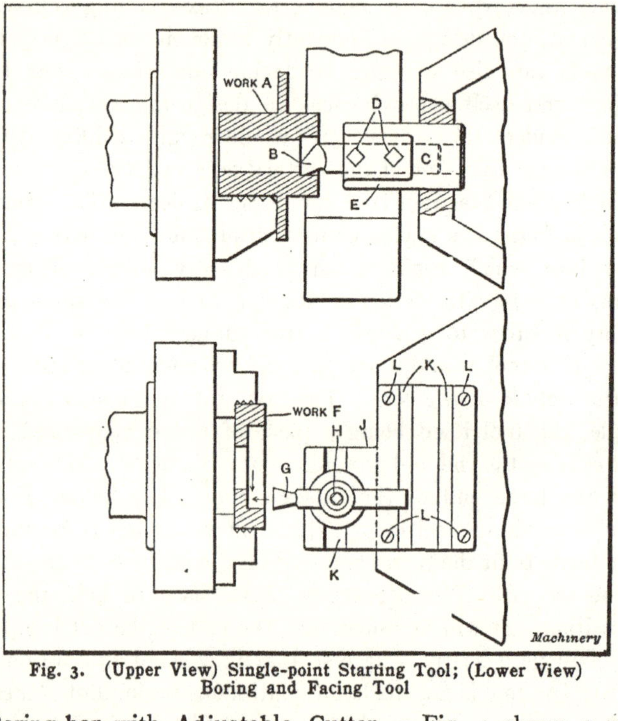

Fig. 3. (Upper View) Single-point Starting Tool; (Lower View) Boring and Facing Tool

Boring-bar with Adjustable Cutter. — Fig. 4 shows a very simple type of single-point adjustable boring-bar for machining the bushing *A* (see upper part of the illustration). Although this bar is simple in its construction, there are several important points in design which should be carefully noted. The bar *E* is of a low grade of tool steel and is hardened and ground to fit the turret hole. The reason for making the bar of tool steel is simply to obtain all the rigidity possible and thereby obviate chatter. The tool B is of round section and is put through the bar at a slight angle, being held in position by the two screws C. A backing-up screw D permits careful adjustment to be made. The bar is cut away where the tool comes through to provide chip clearance, but it is cylindrical on the end except in this one place. By making it this way, it is found an easy matter to use micrometer calipers across the bar and tool as indicated at F, so that accurate settings may be readily made without resorting to "cut-and-try" methods. It is very bad practice to bevel the end of the bar at G and put the holding screws through at this point, because a calipering point is sacrificed by so doing. A simple formula is here given for setting tools of this type for turning a given diameter:

Let F = required calipering distance for a given size hole;
X = diameter of the bar at the end where the tool is;
Y = diameter of the hole to be bored.

Then

$$F = \frac{X + Y}{2}.$$

This formula will be found useful for setting tools very close to the desired diameter, although the spring of the bar will cause slight variations and the amount of stock which is to be removed also makes some difference.

Boring-bar with Double-ended Cutter.—The lower part of Fig. 4 shows a boring tool of an entirely different type. The cutter is double-ended, and a bar of this sort is often used for removing stock rapidly. Although it is a faster cutting tool than a bar having only a single tool or cutting point, it cannot be depended upon to produce a hole which is absolutely concentric with other surfaces machined at the same setting. The work A is the same as in the upper part of the illustration, and the bar B fits the turret hole. It is flattened slightly on two sides at points D, and a rectangular slot contains the cutter C and the wedge E. It will be noted that the cutter is shouldered so that it is a close fit at the points *D*. Tools of this type cannot be ground radially without changing their diameters, but this is seldom necessary as the cutting edge is at the forward end. A land of about $\frac{1}{8}$ inch is usually left just back of the bevel, and the cutter can be ground back to this point without changing the diameter. Beyond this, however, there is a slight back taper for the sake of clearance, so that the life of the cutter does not extend beyond it.

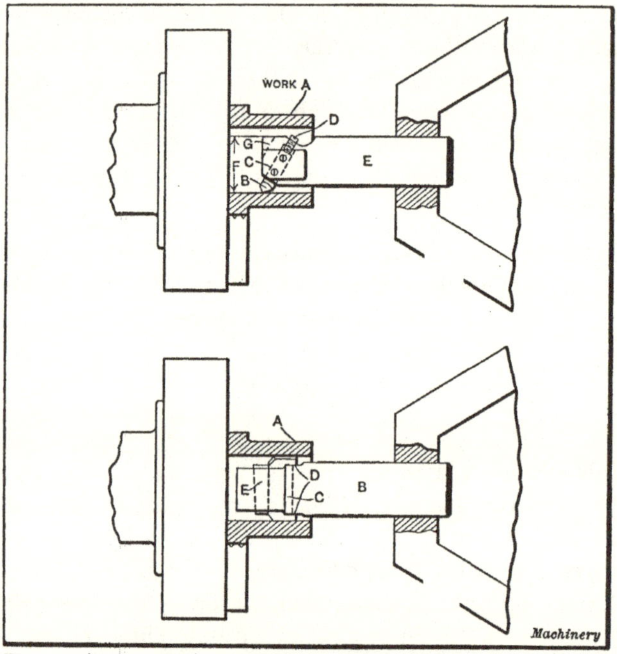

Fig. 4. (Upper View) Boring-bar with Adjustable Cutter; (Lower View) Boring-bar with Double-ended Cutter

Boring-bars with Provision for Cutting Lubricant. — On certain classes of work it is very difficult to supply the cutting points of the tools with sufficient lubrication to make them thoroughly efficient, when the regular supply system is used. Some method must be devised, therefore, to direct the flow to the point or points where the cutting action takes place. An example of a bar arranged to carry the lubricant to inaccessible tools is shown in the upper part of Fig. 5. The work *A* in the chuck jaws is an automobile hub of malleable iron. It will be noted that the portion bored by the forward tool *C* is in such a position that it cannot be reached for lubricating purposes in the ordinary way, but the rear tool *D* can easily be taken care of. The boring-bar *B* is of a low grade of tool steel and fits the turret hole at the rear end; the forward end *J* is a running fit in the chuck bushing *L*. A telescoping oil-supply tube *K* enters this end of the bar and is supplied with

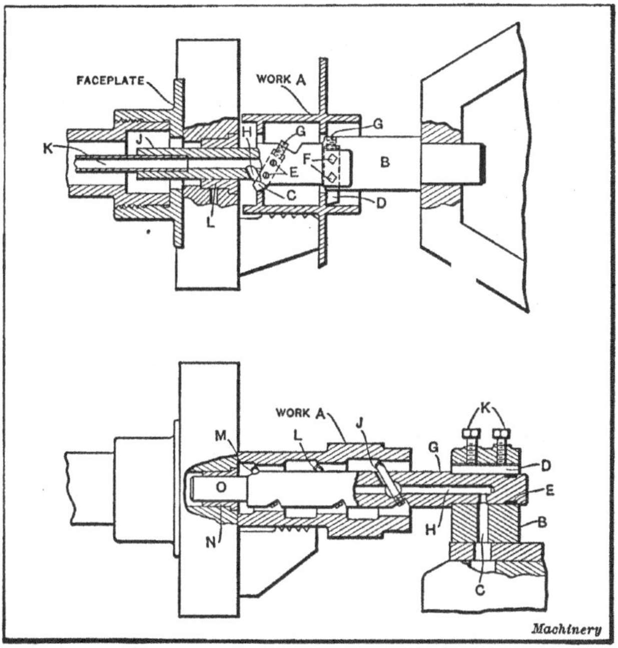

Fig. 5. Boring-bars arranged for Lubricating Cutters

lubricant from the rear end of the spindle. The hole in the bar at *H* leads the fluid directly onto the face of the cutting tool *C*, thus insuring constant lubrication at this point. The two tools are held in place by the screws *E* and *F*, and they are provided with means of adjustment in the backing-up screws *G*. The author has used bars of this type in a number of cases with very gratifying results.

Fig. 5 (lower illustration) shows a very different condition, in a multi-cutting boring-bar for generating a series of true holes in the bronze artillery hub *A*. The finished hole is tapered but a starting bar was used in order to prepare the hole properly for the taper tools which followed it, so that they would not be influenced by the irregularities of the cored hole. In this case the turret lathe was one of the flat-turret variety, and provision was made for lubrication through the hole *C* in the turret face. As the turret indexed to the proper position, this hole came directly over another in the slide, which, in turn, was connected with the lubricant pressure supply system, thus allowing the liquid to pass up into the body of the tool-holder. The boring-bar *G* is turned down at the rear end to fit the tool-holder *B*, and has an annular groove *E* which is packed with felt to prevent the escape of lubricant. A shoe *D* is forced down on the bar by the two screws *K* and prevents the bar from turning. The hole *H* in the bar is drilled from the forward end and is tightly plugged so that this end remains closed to prevent the lubricant from passing through. A groove is cut in front of the tools *J*, *M* and *L*, as shown at *J*, and this allows the fluid to flow directly onto the faces of the tools. The end of the bar is piloted at *O* in the bushing *N* which is fixed in the chuck body. An arrangement of this sort has also proved successful in a number of instances.

Bar for Undercutting, Facing and Boring in the Vertical Turret Lathe. — A very difficult condition for which to design tools is shown in Fig. 6, as the work itself requires rapidity of handling and is a steel casting weighing about 300 pounds. Only a part of the piece is shown at *A*, but it will readily be seen that it is necessary to make the bar in such a way that the tools can be used to do all the cutting indicated by the arrows; *i.e.*, undercut the upper flange, double-bore the interior and face the lower shoulder. As the fixture on which the work was held was of the indexing variety and was very much off center, it was not expedient to run at high speed. Therefore, the double boring was of assistance in increasing the production. It will be noted that the hole through which the tools pass is of small diameter, which makes the problem still more difficult. The shank of the bar *B* fits the turret hole at its upper end and is slotted so that the pin *F* in the turret will act as a driver. (This feature is patented by the Bullard Machine Tool Co.) The lower part of the bar is eccentric to the shank in order to obtain the necessary clearance when the tools are in action. Even the tools themselves are considerably out of the ordinary in that they will cut in two directions. The upper tool *D* is used for undercutting the flange and also for boring, while the lower tool *E* is used for facing the lower shoulder and partially boring the interior. Both these tools have backing-up screws *G* and are held in place by the headless set-screws.

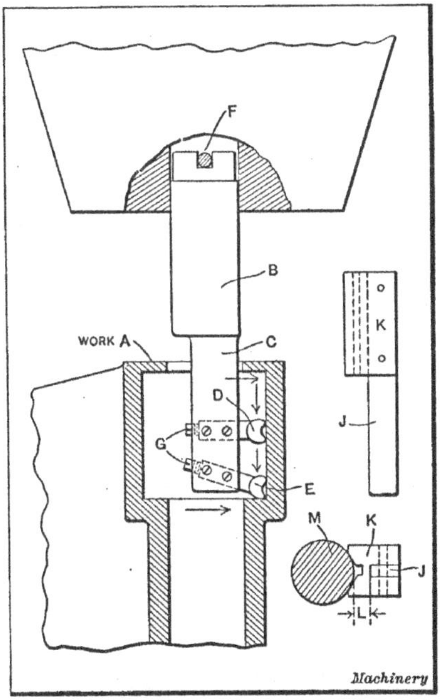

Fig. 6. Bar for Undercutting, Facing and Boring in the Vertical Turret Lathe and Gage used in Setting Tools

As it was necessary to set these two tools so that they would cut approximately the same diameter, the gage shown at the right of the figure was made to assist in the setting. The V-block *K* was slotted to receive the steel strip *J* so that distance *L* would measure the correct distance from the bar shown in section at *M*. It is obvious that the gage could be placed against the bar so that tools could be set out the right amount by means of the backing-up screws. This bar gave fairly satisfactory results although some trouble was experienced with chips, as there was considerable stock to remove. There was likewise a slight tendency to chatter when using a heavy feed, but this was remedied by careful grinding to make the cut as easy as possible. It must be remembered that the conditions were about as awkward as they could be, and the lack of room made it necessary to cut down the bar to such an extent that it was hardly heavy enough for the work. Taken as a whole, however, the action was satisfactory for a roughing tool. It was not used for finishing cuts.

Slip-cutter Bar for the Vertical Turret Lathe. — The steel hub shown at *A* in Fig. 7 is held on a special fixture, located by the previously bored and reamed hole which fits the stud *H*. The hole *K* has been rough-bored in the first operation, but enough stock has been left for the final finishing so that it may be finish-bored and reamed and part *L* threaded at the same setting. This type of bar is the product of the Bullard Machine Tool Co., and is designed especially for use in their machines. It is a combination boring- and reaming-bar, and the cutters are of the "slip" variety. One bar can be furnished with a set of cutters for the various sizes of holes within its capacity. A full set of cutters for any one size of hole consists of chamfering, rough-boring, finish-boring, rough-reaming and finish reaming tools. The first three of these are of square section, carefully ground to fit the broached hole *D*. The rear ends of these tools bring up against the shoulder of the screw *E*, which acts as a stop. The fit in the hole is such that tools can easily be put in and taken out with the fingers.

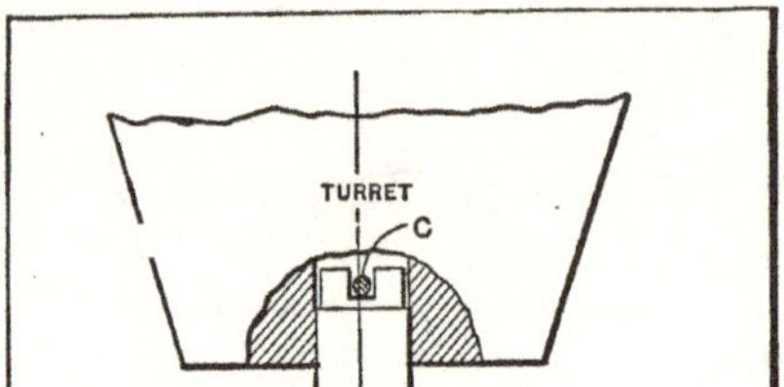

Fig. 7. Bar equipped with a Set of Interchangeable Cutters for Boring and Reaming

The two reaming cutters are used in the rectangular hole F which is at right angles to the other hole; these tools are allowed to float so that they follow the hole generated by the boring tools. The action against the reaming cutters is in an upward direction, and comes against the hardened steel plug *G* which is inserted in the bar. The bar *B* is of special steel and is slotted at the upper end to fit the driving pin *C* which is located in the turret. Bars of this type have a number of advantages, one of which is that only one turret hole is occupied; other advantages are the cost of maintenance, and the adaptability of the bar with its series of cutters to handle a number of different sized holes. The cost of large sizes of reamers of the fluted type is considerable, while a flat reamer, such as that used in this bar, is inexpensive. It may be noted that the pressure or thrust of the cut is all that holds the boring tools in place, so that trouble would be experienced if a cored pocket were to be encountered. This is provided for by a detent screw in the end of the bar, which prevents the tools from coming out. This screw can be put in any time if it is found necessary.

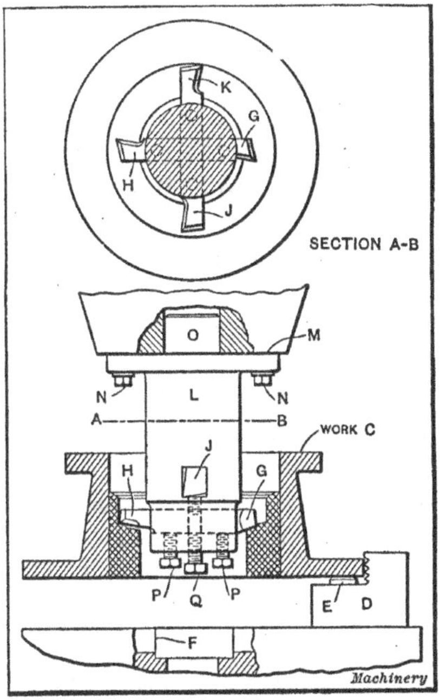

Fig. 8. Bar with Four Cutters set at Different Radii for Removing Considerable Stock in One Cut

Multi-cutting Bar for the Vertical Turret Lathe. — An example of a bar designed to remove a large amount of stock in a very short time is shown in Fig. 8. The work *C* is a steel boiler nozzle which is forged, and a 5-inch hole punched in it, before it is machined by the vertical turret lathe. The finished hole is 8 inches in diameter and it was desired to remove the surplus stock as rapidly as possible. Accuracy in the diameter of the hole was not essential. The bar *L* is a steel forging which fits the turret face at *M* and is held to it by the screws *N*. The stem *O* is used to center the bar in the turret. Two rectangular holes are broached through the body of the bar, at right angles to each other, and these receive the high-speed steel cutters shown. It will be noted that these cutters are so proportioned that they remove the stock in a series of steps, each tool extending beyond the preceding one and also above it about $\frac{1}{8}$ inch.

The section taken at *A–B* shows the manner in which the tools extend beyond each other, and the lower view illustrates the cutting action of the tools. One end of the first tool *G* makes the first step, while the other end *H* makes the second. In like manner one end of the upper tool *J* makes the third step while the other end *K* takes out the remainder. The two screws *P* hold the first tool in place, while the other is secured by the screws *Q*. It will be noted that the work is held in jaws *D* and is supported on buttons at *E*; the height above the table is great enough to allow the end of the bar and the set-screws to go through far enough to finish the cut. Regarding the upkeep of these tools, attention is called to the fact that they may be pushed backward or forward to compensate for wear and distribute the cut. For roughing purposes requiring rapid removal of stock and rapid cutting, a bar of this sort has proved very successful, but it is not recommended for work requiring great accuracy.

In several of the illustrations it may be noted that the tools are shown in a plane parallel with the rotation of the turret. This has been done simply because the details are more clearly apparent when shown in this way. Greater accuracy is obtained by setting the tools in a plane perpendicular to the turret rotation, as previously stated.

CHAPTER III

RECESSING TOOLS

Kinds of Work Done by Recessing Tools. — Many varieties of cylindrical work call for the machining of an annular recess or groove in a place which may be inaccessible to the cutting tools. The form of recess varies greatly and the accuracy required is likewise variable. The form may be either narrow or wide, deep or shallow, while the accuracy called for may be either within narrow or liberal limits, as, for instance, when the recess is for clearance only. In fact, in the majority of cases the purpose of the relief or recess is merely to obtain clearance for some moving part or for tools when machining an adjacent surface. Very frequently a groove is cut to serve as an oil pocket or to provide a space which can be filled with packing to act as a gland. It is evident that great accuracy is not essential when the work is of this nature. There are occasionally conditions which require more accurate work, as, for instance, when another piece is to be sprung into place, such as a spring ring or something of a similar nature, but even in a case of this kind a certain amount of inaccuracy is permissible. The machines to which recessing tools are most frequently fitted are the engine lathe, the horizontal turret lathe, the vertical turret lathe, the vertical drilling machine and the horizontal boring mill. Other machines are occasionally equipped with tools for the same purpose, but those mentioned are most frequently used.

In many cases the position of the relief or groove is such that it cannot be readily seen by the operator, nor can it be easily calipered. The workman, therefore, must tell how the tool is cutting by the "feeling" of it and by the character of the chips. He is really "working in the dark," and for that very reason every precaution must be taken, in regard to position of tools, diameter and shoulder stops, etc., so that the machining can be done without withdrawing the tool to note the progress of the work. In this connection it is well to bear in mind that the action of any kind of grooving tool is much the same as a cutting-off tool. It must be kept very sharp and set so that the cutting edge is slightly above center, when it is used for internal work. It will be seen that if the tool is slightly above center the springing down of the cutting edge (due to the pressure of the cut) will have a tendency to keep it from "digging in," and will therefore assist in the prevention of chatter.

Points in Design of Recessing Tools. — Some of the important points in the design of recessing tools are as follows:

1. Rigidity is of the greatest importance and every precaution should be taken to insure as substantial a holder as possible. The tool itself should be of as great a section as the conditions and the space will permit. Some method of supporting the overhanging end should be provided, either by means of a pilot or in some other way which may suggest itself. Moving parts should have a means of adjustment for wear, and gibs should be set up as snugly as possible and still allow free movement.

2. The feed motion should be carefully considered. Screw feed is best, and may be contained in the tool itself or may be operated by the cut-off slide. Lever feed is uncertain and produces uneven cutting unless the work upon which it is used runs at high speed. When this is the case and if the cut is not too heavy, it can be used with satisfactory results. The work to be done is a factor in determining the method most satisfactory for the feed motion.

3. Means are needed for determining the depth of the cut. There are several ways in which the depth of the cut can be positively determined: a positive stop can be provided; the dial on the cut-off slide can be used when the feed motion of the slide is the operating force; an indicator or a graduated dial on the tool-holder itself may be provided.

4. Rapidity of operation is essential.

5. Adjustment for the cutting tool should be provided.

This adjustment may be made either by manipulating the tool by hand or by means of a backing-up screw. The latter method is the better one and should be used whenever practicable. The upkeep of the tool is important, and for that reason inserted

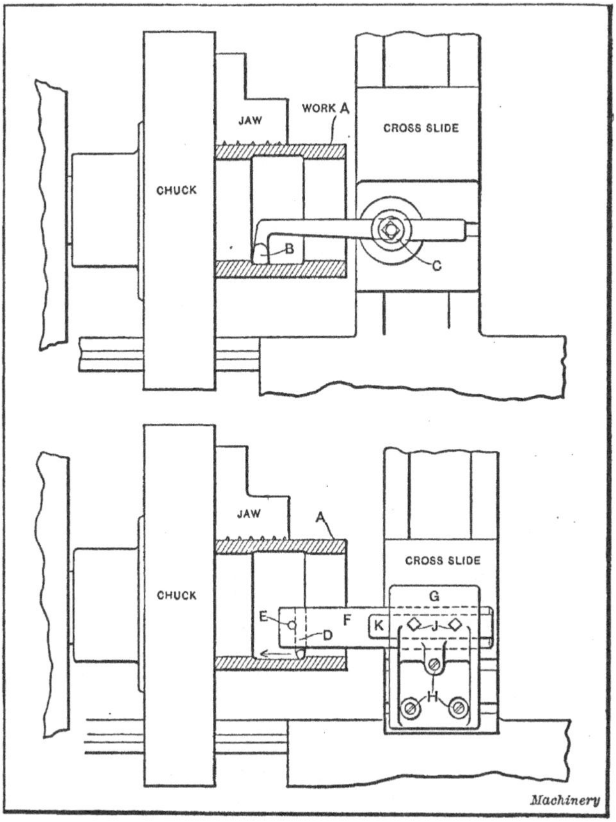

Fig. 1. Two Simple Types of Recessing Tools for the Engine Lathe

tools are preferable to those which form a part of the mechanism itself. In confined situations it is occasionally necessary to make the tool of special shape. This should be done only as a last resort, when necessitated by the conditions governing the work. In cases of this kind several tools should be made to provide for emergencies.

Recessing Tools for the Engine Lathe. — The upper illustration in Fig. 1 shows a bushing *A* which is held by the outside in regular chuck jaws. This work is to be done on the engine lathe, and the recess is to be cut at the same setting. A forged tool *B* is held in the regular toolpost *C* on the cross-slide

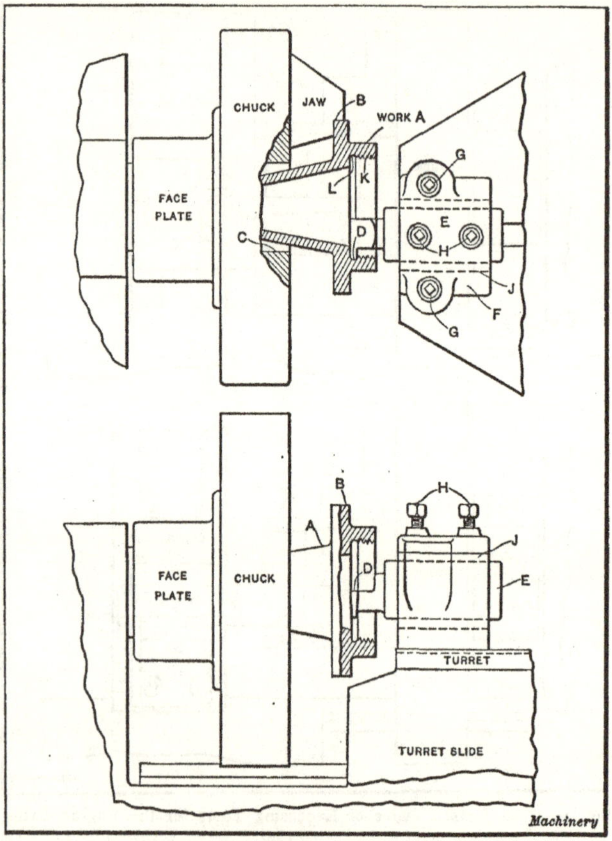

Fig. 2. Recessing Tool used in a Turret Lathe

of the lathe, and is forced into the required depth by hand. After this the longitudinal feed is started and the remainder of the recess cut. This type of tool is much used for lathe work when only one or two pieces are to be machined. Its advantages are that it can be easily made and quickly adjusted.

Its advantage is that it has a tendency to chatter, and is, therefore, suitable only for very light cutting.

The device shown in the lower portion of the same illustration is much more rigid, but is not nearly so adaptable to various conditions. In this arrangement, the tool *D* is of round section and is held in place by taper pin *E*. The bar F is of steel and is secured in the holder *G* by the two screws *J* which bear against a flat *K* on the bar. Three screws *H* enter shoes in the cross-slide T-slots and secure the holder firmly to the slide.

Recessing Tool for a Horizontal Turret Lathe. — The work *A*, shown in Fig. 2, is a steel forging of an automobile hub which is held in a three-jawed chuck by the flange *B*, the tapered portion entering the hole *C* in the chuck body. The inside of the hub is to be threaded at *K* with a collapsing tap. A recess is therefore needed at *L* in order to obtain a clean thread. The machine selected for the work is a Pratt & Whitney turret lathe having a cross-sliding turret of the flat type. The recessing tool is of high-speed steel, with the shank turned and ground cylindrical at *E*. The front end is also turned to form the flange *D*, and is afterward cut away and finished to the shape required, as clearly shown in the lower part of the illustration. The tool-holder *F* is of cast iron and contains a steel split bushing *J* which is compressed by two screws *H* in the top of the holder. The action of this tool was satisfactory, but the upkeep is obviously rather expensive.

Eccentric Recessing Tool for a Horizontal Turret Lathe. — The work *A*, shown in Fig. 3, is a steel flange which is to be recessed at *B* in order to provide the necessary clearance for the threaded portion *C*. In this instance the cut-off slide was used during the progress of the work, so that a considerable overhang from the turret was required. Strictly speaking, this is not an eccentric tool, for the various parts of the body are concentric, but by a reference to the upper part of the illustration it will be seen that the center-line *VW* of the recessing tool does not coincide with the center-line *TU* of the spindle. Now as the tool-holder F revolves on the center-line

VW, it is evident that the path of the tool *D*, as it swings, will be eccentric to the center-line of the spindle. The body *L* is of cast iron and is mounted on the dovetailed turret face, being securely held in position by the gib *M* and the screws *N*.

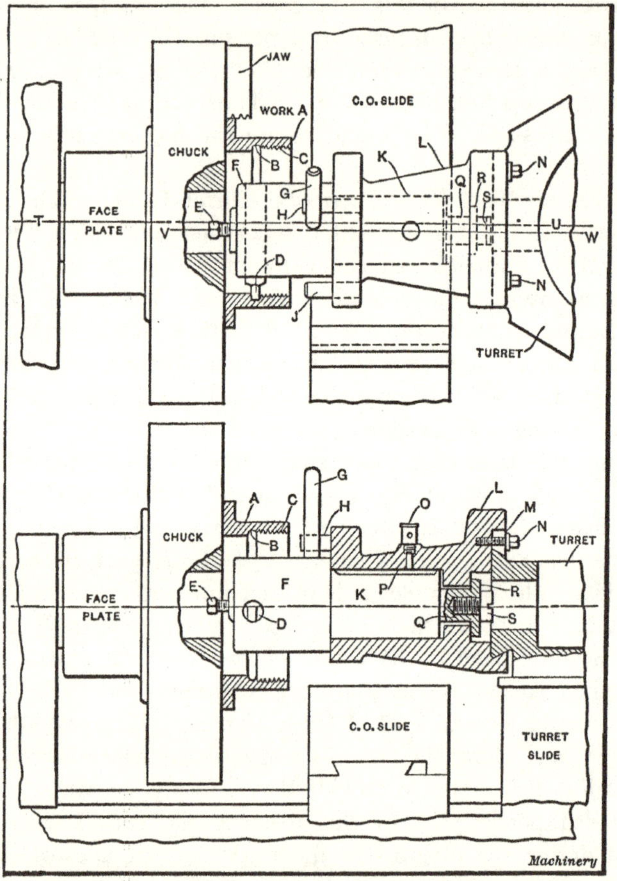

Fig. 3. An Eccentric Recessing Tool for the Turret Lathe

The tool-holder *F* is of tool steel and is turned down at *K* to a running fit in the body. The end *Q*, with the screw and washer *S* and *R*, acts as a retainer to keep the tool-holder in position. The tool *D* is of round section with the cutting end so shaped that it will cut the recess properly. A set-screw *E* holds it in position. An oiler *O* is located in the body and distributes the oil to the bearing through the oil groove *P*. An operating handle *G* is driven into the holder, and is located between the pins *H* and *J* which act as stops. As the lever *G* is operated, the tool *D*, starting with slight clearance at the bottom of the hole, moves gradually upward and outward until the full depth of cut has been reached. At the completion of the cut the tool stands in the position shown in the illustration. The action of this tool was perfectly satisfactory, and as it is comparatively simple in construction, the cost of building was not excessive.

Piloted Recessing Tool for an Automobile Flywheel. — A rather peculiar condition is shown in Fig. 4, the work *A* being an automobile flywheel having a semi-circular recess at *B*. Attention is called to the fact that this recess is put in at an angle of 45 degrees with the center-line. It is evidently only a clearance groove for the male clutch member, and it is not known to the author why some other style of groove would not have answered the purpose just as well.

The work *A* is held by the inside of the rim in special jaws. The body of tool *D* is made of cast steel and is fitted to the dovetailed face of the turret, the gib *G* securing it firmly by means of the collar-head screws *H*. A tool-steel pilot *E* enters the bushing *C* in the chuck and assists in supporting the body against the pressure of the cut. This pilot *E* is shouldered and forced into the body at *N*. A small hole *O* is drilled to avoid air compression when forcing in the pilot. If this is not done, the fitter may be deceived into thinking that he has secured a good fit at this point, when in reality it is the air compression which causes the stem to fit tightly.

A cover plate F tends to strengthen the body and overcome the weakening effect caused by the cutting of the angular slot, and also assists in preventing the entrance of dirt and chips. Tool *J* is of square section and is held in the sliding block *M* by two screws *L*. Hole *K* is for machining purposes only. The operating screw *P* is squared up on one end to receive a wrench, while the other end is shouldered at *R* and threaded to receive a hexagon nut. There are two thrust washers shown at *Q*. The screw has four Acme threads per inch, right-hand, and meshes with the angular rack cut on the under side of

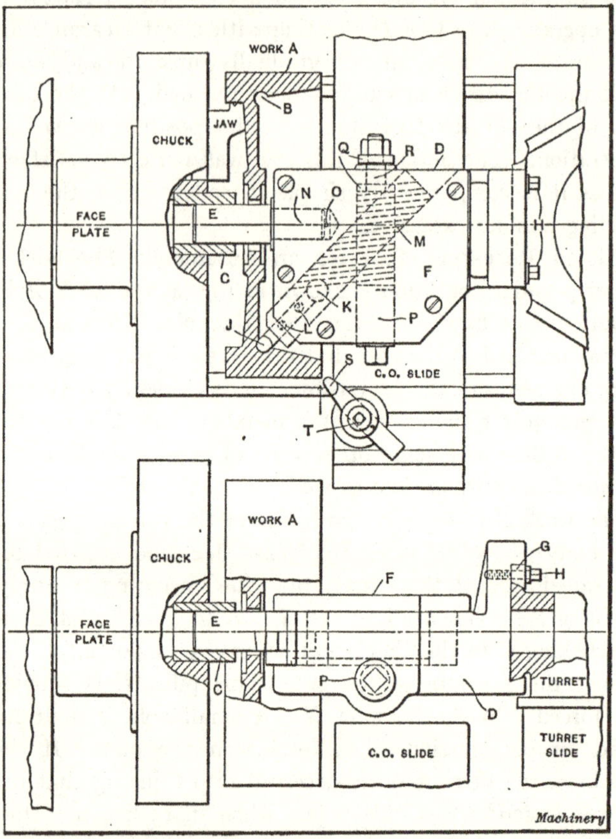

Fig. 4. A Recessing Tool used in Machining an Automobile Flywheel

the tool-carrying slide *M*. It is evident that the rotary motion of screw *P* will cause movement of the block, in its longitudinal direction, thus feeding the tool into the work at the desired angle. The forged tool *S*, held in the tool-holder *T* on the cut-off slide, is slowly fed across the rim while the recessing operation is taking place. This arrangement of the tools gave very satisfactory results.

Double Recessing Bar for a Rear Axle Housing. — The work *A* shown in Fig. 5 is a bronze rear axle housing for an automobile, and the recessing bar is only one of a group of tools

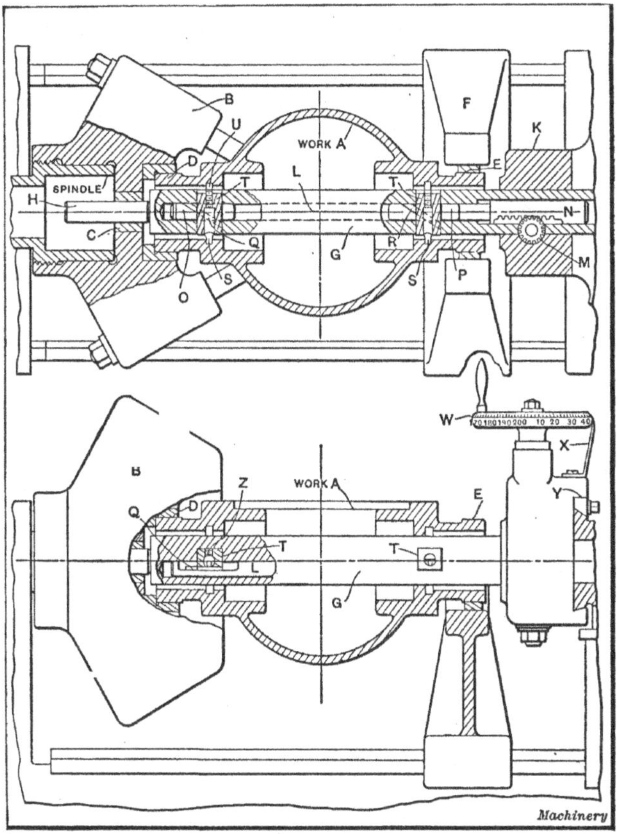

Fig. 5. A Double Recessing Tool Arrangement for a Rear Axle Housing

used at the same setting of the work. Previous to this setting the finished annular rings at the two ends *D* and *E* of the casting were machined so that they might be used as locating points in this setting. The ring *D* slips into the split bushing in the holding device *B*. The other end *E* revolves in a roller back-rest *F* which is placed on the ways of the turret lathe. This back-rest is not shown in detail, as its construction is not essential in connection with the recessing tool. The two grooves in the work at *S* were to be spaced an exact distance apart and it was partly to insure accurate spacing that this bar was designed, although rapidity of operation was also a factor. A cast-iron bracket *K* is fastened to the dovetailed face of the turret by means of gib *Y*, shown in the lower view. The hand-wheel *W* is connected to a shaft which drives the pinion *M*. A steel pointer *X* is fastened to the bracket and acts as an indicator on the graduated rim of the wheel. It will be seen that this arrangement makes it very easy to determine the depth of the cut.

The pinion *M* meshes with a rack cut upon the enlarged end *N* of the operating rod *L*. This rod is considerably below the center of the bar and is flatted at *O* and *P*. The tongues *Q* and *R* are angularly cut on these surfaces, and they engage with grooves on the under side of the tool-carrying blocks *T*, so that any longitudinal movement of the rod *L* is transformed into a radial movement of the blocks. The grooving tools *S* are of round section and are held in position by the headless screws *Z*. The backing-up screws *U* permit accurate adjustments to be made with ease. The pilot *H* enters the steel bushing *C* in the body of the holding device and assists in preventing chatter. An added refinement to this tool was an oil groove from which oil was led directly to the cutting tools. This was supplied with oil through a special piping system and a distributing collar on the turret. In order to avoid confusion in the drawing, this has not been shown. This device gave very satisfactory results.

Recessing Tool for an Automobile Bearing Retainer. — The work shown at *A* in Fig. 6 is a malleable iron bearing retainer for an automobile. The casting is held by the outside in a three-jawed chuck; the machine on which the operations are performed is a horizontal turret lathe. The piece is completely finished in one setting. As the cut-off slide front tool carrier was used during the progress of the work, it was found necessary to design the recessing tool so that it extended out over the slide. It is evident that an overhang as great as this would cause trouble unless some means of intermediate support were provided. The bracket *S* was therefore used on the rear of

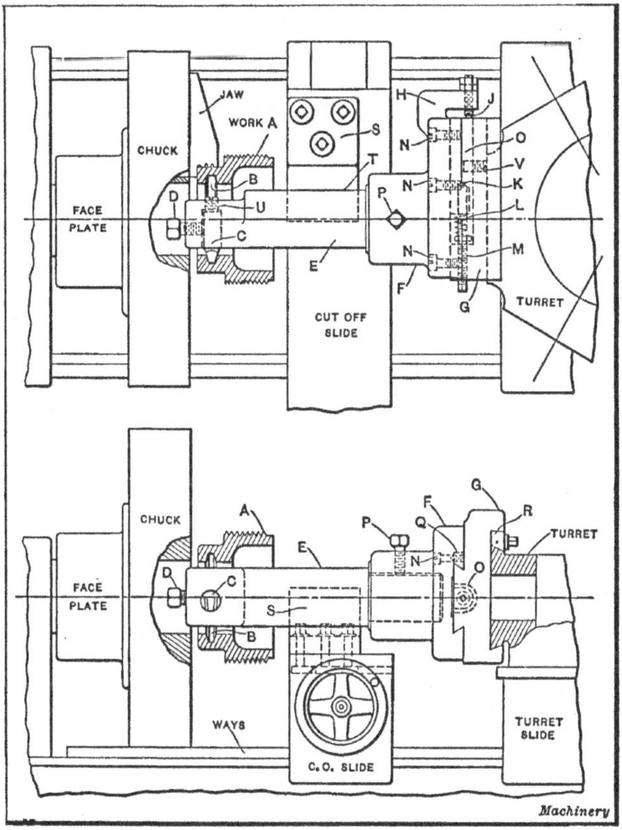

Fig. 6. A Recessing Tool for an Automobile Bearing Retainer

the cut-off slide, the portion *T* being cut out to the radius of the bar so as to act as a support and at the same time provide the feed motion necessary (through the reverse feed of the slide) to force the tool into the work.

The cutting tool *C* is of round section properly shaped at the end to form the required groove *B*. It is secured in place in the bar *E* by the set-screw *D*; radial adjustment is secured through screw *U*. The rear end of the bar is shouldered and fitted to the sliding bracket F; the set-screw *P* holds it in place. The slide *F* is dovetailed and is gibbed to the fixed bracket *G* by the gib *Q* which is adjusted for wear by the screws *N*. The lug *H* at the end of the slide is provided with a stop-screw *J* which permits close adjustments to be made for the depth of cut. This lug is not shown in the lower view, but it is set slightly to one side of the cored groove *O* so that the screw will bear against the solid portion. The bracket *G* is mounted on the dovetail of the turret and is held in place by the gib *R*. The special screw *M* is shouldered to receive the coil spring *L* which thrusts against it and against lug *K* on the slide. The strength of the spring may be easily adjusted by the screw to the desired compression. The screw *V* is simply used to limit the reverse movement of the slide, so that it will not move back too far before or after the work has been done. This device was used for three different pieces by simply changing the tool and regulating the stop-screw. Its performance was thoroughly satisfactory.

Recessing Tool for a Large Collar. — The large collar *A* in Fig. 7 was held by the outside of the flange in a three-jawed chuck on a horizontal turret lathe. The internal groove *D* was to be cut during this setting of the work, and as a small geared scroll chuck was conveniently available, it was arranged as a recessing device for this casting. The cast-iron bracket *H* was fitted to the faceplate recess at the rear of the chuck body. The stem *J* was turned down to fit the hole in the turret face, and the four screws *K* secured it thereto. One of the standard chuck jaws F was annealed and shaped up as shown. It was then drilled to receive the tool *E*, and tapped out so that two set-screws could be used to hold the tool. The jaw was then rehardened and a small amount of fitting done so that it worked smoothly. A graduated collar was applied at *M*, and a special wrench *L*, having a slip handle, served to operate the scroll and thereby caused the tool to move radially as required. A tool-steel pilot *C* was forced into the center hole in the chuck body *G*, and a bushing *B* in the spindle chuck body served as a guide and support for it, thereby greatly increasing the efficiency of the tool and doing away with the chance for chatter.

Recessing Bar for a Triple Groove. — In all of the examples which have so far been given, the work has been done in a

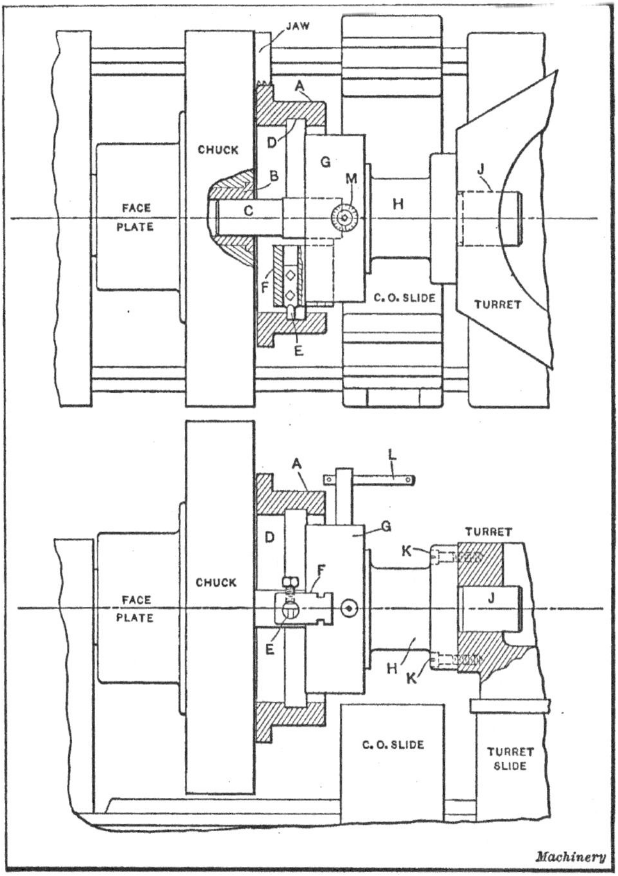

Fig. 7. A Recessing Tool for a Large Collar, used in a Turret Lathe

horizontal plane, but we shall describe a few cases which are handled in a vertical plane on the vertical turret lathe. As this machine has a turret slide which can be traversed horizontally, it is evident that no special attachments are required for plain recessing or grooving, but there are conditions which may be decidedly out of the ordinary, under which a special

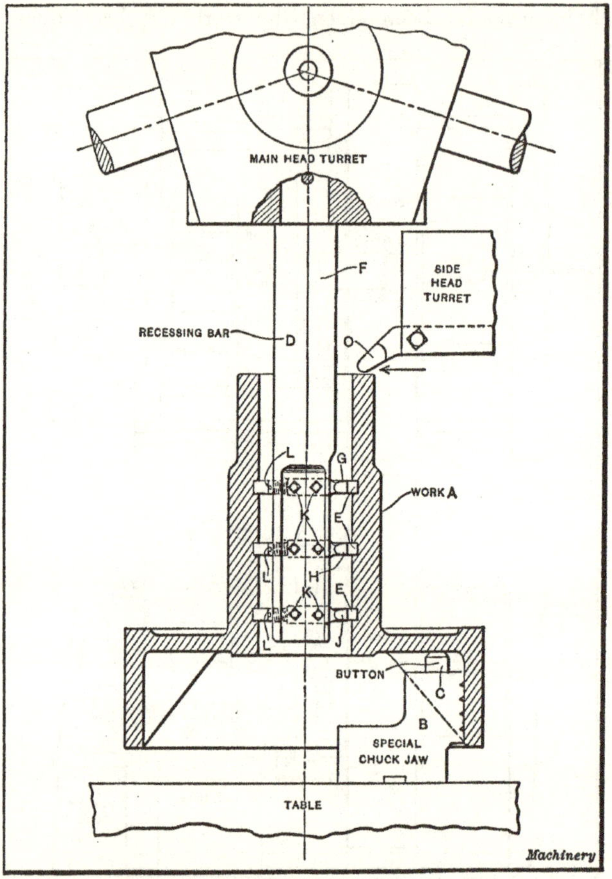

Fig. 8. A Multiple Recessing Tool used in a Vertical Turret Lathe

arrangement for recessing may be used to advantage; for example, any sort of groove which is deep down in a hole, multiple grooving at a considerable depth, or any other condition of a similar nature.

When the groove is very deep there is naturally a considerable overhang of any tool which may be used for the work. If the overhang is excessive, it follows that there is apt to be more or less vibration, and vibration means chatter. If, however, a tool or bar having an excessive overhang from the turret is supported at its lower end, the tendency to chatter is at once overcome; but, if support is provided at this point, the horizontal movement of the turret slide cannot be used. Therefore, some method which will give a radial movement to the grooving tool must be used when the bar is to be supported at its lower end.

Fig. 8 shows a piece of work at *A* which is set up so that it can be machined complete in one setting. The casting is held by the inside of the rim in special chuck jaws *B*, and is supported at three points on the steel buttons *C* which rest in pockets in the jaws. The inner ribs of the casting act as drivers against the sides of the jaws. The three grooves *E* are to be machined and the tools *G*, *H* and *J* are correctly spaced to perform the work. These are secured in the bar *D* by means of the set-screws *K*, and accurate adjustment is provided by screws *L*. The bar *D* is shouldered at the turret face and is driven by a pin in the turret in the usual manner. The tool *O* in the side-head turret is used for facing while the inside work is being done, as this brings the cutting action of the outside and inside tools in opposition and therefore tends to overcome vibration. If a very fine feed is used on the turret traverse, good results may be obtained with this method, although there is a tendency to chatter due to the excessive overhang. Slight variations in the depth of the grooves may also be found on account of the spring of the bar.

Piloted Recessing Bar for a Triple Internal Groove. — The cast-iron valve cap shown in Fig. 9 is another example of a piece of work having three grooves equally spaced, and in which the lower groove is at a considerable distance from the turret. This piece is finished complete in one setting and is held by the outside of the flange in the standard chuck jaws *C*, being supported at three points by the buttons *D*. This tool

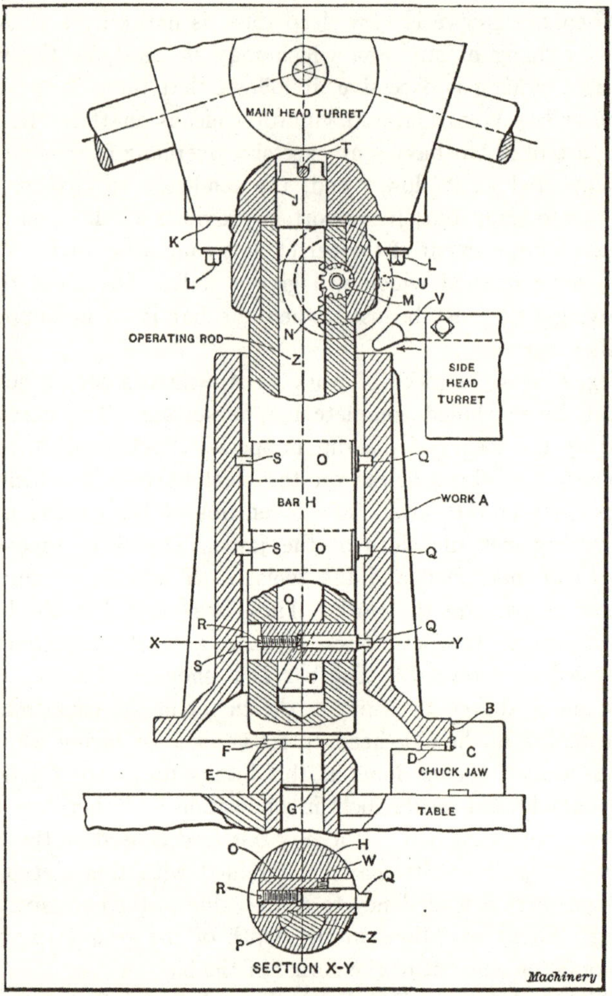

Fig. 9. Another Multiple Recessing Tool used in a Vertical Turret Lathe

is somewhat similar in its operation to that shown in Fig. 5, except that it is arranged in a vertical instead of in a horizontal plane. A heavy cast-iron bracket is bolted against the turret face *K* by screws *L*, and a locating plug *J* centers the device in the turret. The bar *H* is of steel and has a pilot *G* at its lower end. This pilot is hardened and ground to fit the bushing *E* which is inserted in the center of the table. The top of the bushing is milled out to leave three projecting pads *F*. These pads form a positive stop to insure the correct height; it will be noted that the tendency, when in action, would be to keep these pads clean and free from chips or dirt. The upper end of the bar is shouldered and is fastened to the bracket. As in the former instance the operating rod *Z* is flatted at certain places and angular tongues *P* are provided. These tongues mesh with corresponding grooves in the tool-carrying blocks *O*. The section *X–Y* gives a good idea of the construction.

The tools *Q* are held in place by the short set-screws *W* in the square steel blocks *O*. The backing-up screws *R* permit of rapid and easy adjustment. At the upper end of the operating rod the rack *N* is cut and the pinion *M* meshes with it and operates the rod. The handwheel through which the pinion is operated is indicated at *U* by the dotted lines. This portion of the mechanism is identical with that described in Fig. 5. The tool *V* in the side-head turret is used for facing the end of the casting during the progress of the recessing operation.

Recessing Tool Operated by Bevel Gears. — A somewhat unusual condition is shown in Fig. 10, this arrangement having been suggested for the work *A* in order to rapidly perform the grooving operation deep down in the interior of the casting at *V*. It was desired to machine this casting complete at one setting. The chuck jaws *B* were of special form, having a slight angle on the inside of the jaw which drew the casting down onto the three points *C*. A cast-iron pot *E* was fastened to the table by screws *K*, and cored openings *J* were left at the points where the jaws gripped the work. Midway between the jaws, the pot casting took the form shown at *D* and the dogs F were sunk into the edges of the flange by means of the hollow set-screws *G*. The bar *M* is a steel casting which bolts against the turret face at its upper end; it is located by the plug *L*. The operating sleeve *T* is of tool steel, hardened and ground, and having an angular slot *X* at its lower end, which

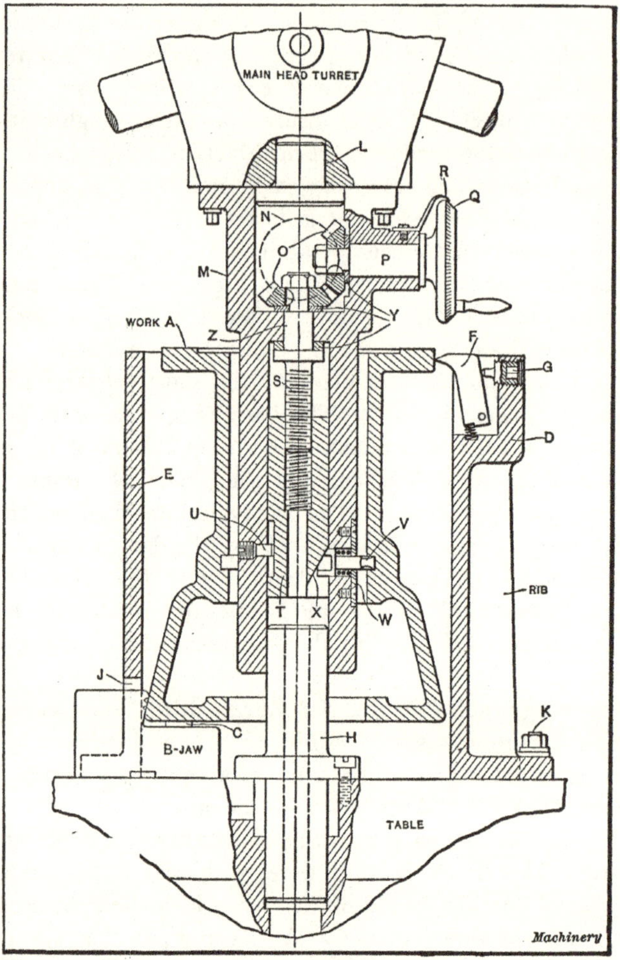

Fig. 10. A Tool for Recessing in a Difficult Position, in Use in a Vertical Turret Lathe

bears against the tool *V*. It is well to make up several of these tools, so that replacements can be quickly made in case of breakage. A steel plate *W* is let into the casting at this point to form a cover plate for the tool and spring pocket. A teat-screw *U* fits a slot in the operating sleeve and prevents it from turning.

The left-hand threaded shaft *S* is journaled at its upper end *Z* and the miter gear *O* is keyed in place. The shaft *P* carries another gear which meshes with the former, and the entire mechanism is operated by the handwheel *Q*. (This handwheel, in reality, is located 45 degrees toward the front of the machine

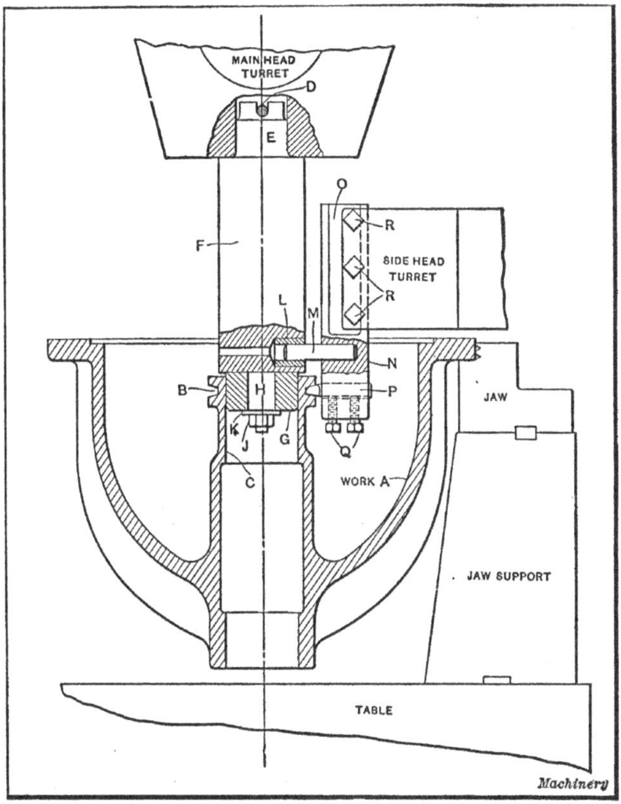

Fig. 11. An Arrangement for cutting a Groove on the Outside of a Sleeve

from the position shown.) A pointer *R* assists in making accurate readings from the graduated bevel on the handwheel. Steel thrust collars *Y* are provided for wear. The tool-steel pintle *H* is fitted to the center of the table and is held down by the screws shown. This pintle acts as a guide upon which the mechanism is located and greatly assists in making it rigid.

Arrangement for External Grooving.—A thin piece of work for electrical machinery, shown at *A* in Fig. 11, has been completely machined with the exception of the groove *B*. At the time when the operation of grooving takes place, a revolving

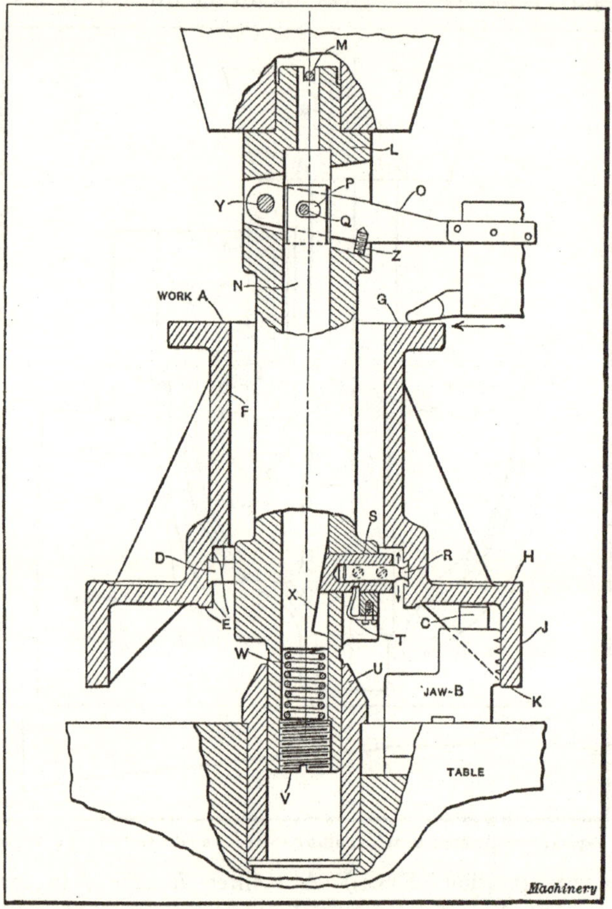

Fig. 12. A Recessing Tool Cutting a Dovetail-shaped Recess

steel pilot *G* fits the previously reamed hole *C* and is held in its position on *H* by the nut and washer *J* and *K*. The upper portion of the bar *F* is shouldered at *E* and fits the turret hole, being kept from turning by pin *D*. A round bar *N* is flatted on two sides at *O* and is held in the side-head turret by the three screws *R*. The lower portion of the bar carries the grooving tool *P* which is held in place by the two screws *Q*. A tool-steel pin *M* is forced into the bar *N* and forms a sliding tie between the pilot bar F and the side-head bar *N*. The bushing *L* is inserted in the pilot bar to receive the pin. It will be readily seen that this method overcomes the vibration which would naturally be caused by the grooving tool acting on the thin and unsupported hub.

Recessing Tool for a Dovetail. — The casting shown at *A* in Fig. 12 is a portion of the clutch mechanism for a farm engine, and it was desired to machine the piece complete in one setting. The work was, therefore, held by the inside by special jaws *B* and supported at three points by the steel buttons *C*. A good driving action was provided by the side of the jaws bearing against the ribs shown. In order to properly make the dovetail cut *D*, it was necessary to move the tool radially to secure the proper depth and then move it upward and downward in order to machine the corners of the dovetail. The shoulder in the casting made it impossible to see the work which was being done. The steel bushing *U* was centered in the table and acted as a guide for the pilot-end of the bar *L*. The entire bar is a steel forging and contains a plug *V* at its lower end against which the coil spring *W* thrusts. The operating rod *N* bears against this spring which is sufficiently strong to keep it up to the limit of its upper movement. The lower portion of the rod is slabbed off at *X* to a 20-degree angle, this angular portion acting as a wedge to force the tool-block *S* outward. A flat spring *T* keeps the tool-block back against the angle on the operating rod and assists in releasing the tool after the groove has been cut. The dovetail tool *R* is held in position in the block by means of the two screws shown. The operating lever *O* is pivoted at *Y* and passes through a slot in the operating rod *N*. A pin *Q* bears against the elongated slot *P* and thereby moves the rod in a vertical direction. The adjustable stop-screws *Z* limit the movement. The lever *O* is shown in a position 45 degrees toward the rear of the machine from that in which it is really located. It will be noted that the side-head tool may be facing the work at *G* during the progress of the recessing operation. In machining this piece the surfaces *D*, *E*, F, *G*, *H*, *J* and *K* were all finished at this setting.

CHAPTER IV

FLOATING REAMER HOLDERS

Requirements of Reamer Holders. — At first glance nothing is more simple than a reamer holder which is so made as to permit the reamer to float in any direction. Yet, if this is so simple, why are there so many types of so-called "floating holders?" All of these must have some special claim to efficiency as a reason for their existence. In order to properly discuss the question, we must first see clearly that there is a necessity for a holder of this type, by analyzing the conditions governing its use, taking as an example a piece of work to be machined in a horizontal turret lathe.

Let us assume that the hole is first carefully generated with a single-point boring-tool held in the turret. It should then be absolutely concentric about the axis of rotation and ready for the reamer. If the reamer is so held that its centers are coincident with this axis and is fed into the work in this position, the resulting hole should be true and of the correct size. Unfortunately, however, machines of this type are subject to slight variations in their indexing, and the inevitable wear upon the ways produces a certain amount of inaccuracy which must be taken care of in some way in the reamer holder, in order that the reamer may be unrestrained in following the generated hole produced by the boring tool. It is now apparent that there is a very good reason for the floating type of holder for reamers used in turret lathe work. The problem consists in selecting from the many varieties a holder which will allow the reamer to follow the generated hole without restraint, even when the turret holes are considerably out of alignment with the spindle.

Types of Reamer Holders. — Let us first consider the holder shown in Fig. 1 — a very common type — consisting of the body

A, the shank of which fits the turret hole, and a stem *B* having a " ball end " through which passes the driving pin *C*. This pin also serves to take the thrust of the cut. The forward end of the stem is tapered to permit the use of various sizes of shell reamers, such as the one shown at *D*. While this holder is probably more commonly used than any other type, its principal claim to efficiency is based on the floating action permitted

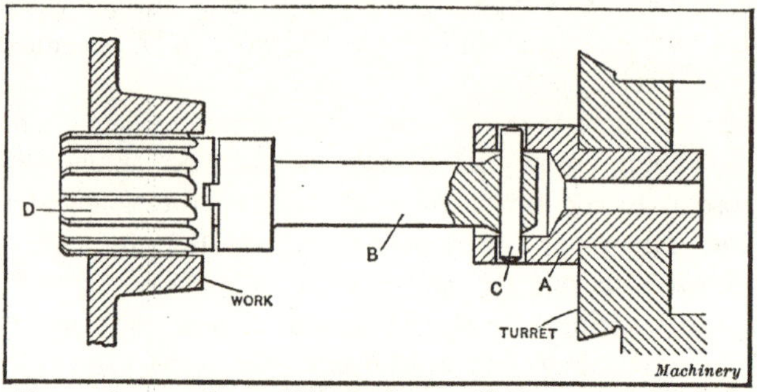

Fig. 1. Common Form of Floating Reamer Holder supported by Ball and Pin

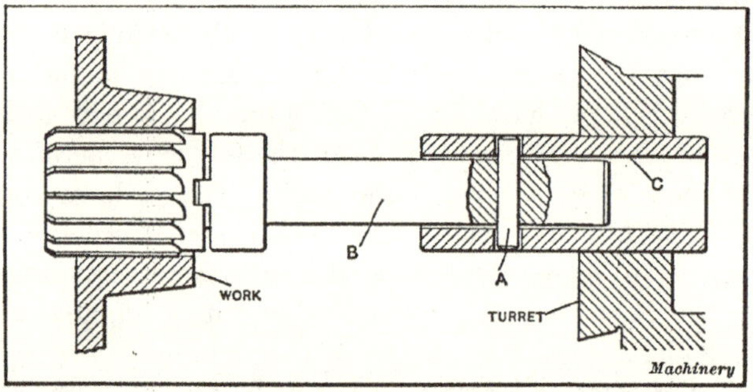

Fig. 2. Another Common Form of Floating Reamer Holder

by the ball end. Yet this ball, although it allows a swivel action, is also a disadvantage in that the reamer hangs so low that it must be very carefully started into the hole by hand. It also has a tendency to produce a slightly bell-mouthed hole, on account of the inevitable wobble when starting. If the ball end is made considerably smaller than the hole in the body (say 0.015 to 0.020 inch) it may be made to produce accurate results when properly handled. Adaptability and moderate cost are points in its favor.

Fig. 2 represents another very common type, in which the driving pin *A* is sometimes a free fit in the body and in other cases loose in the shank. This also is a holder which may be cheaply made, but it is subject to some of the same objections as the tool shown in Fig. 1. It is frequently made with too much clearance between the shank *B* and the hole *C*, this being as much as $\frac{1}{16}$ inch in some cases. From 0.005 to 0.008 inch is much better, as the reamer will then enter the hole without assistance, and is less likely to cut large on its entrance.

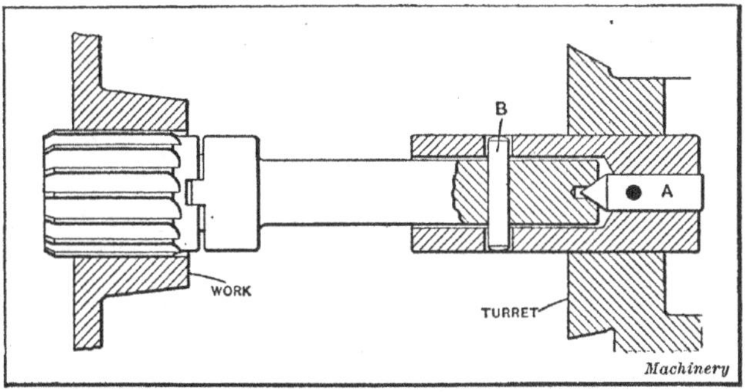

Fig. 3. Improved Form of Holder with Plug A to center the Reamer and take the Thrust

An improved form of this holder is shown in Fig. 3, where it will be seen that a centering plug *A* has been added, which in addition to centering the reamer, also takes the thrust. In certain cases the use of this plug is a disadvantage; for if the turret does not index properly, the plug must obviously be off center, thus producing a like condition in the reamer itself. Fig. 4 shows such a condition in an exaggerated form. The center of the turret hole *A* is 0.010 inch lower than the center of the spindle *B*, due to improper upkeep, excessive wear on the ways, etc. The hole *C* has been generated by a single-point boring tool and, therefore, should be symmetrical about the axis of rotation of the spindle. The reamer itself is 4 inches long and the distance from the forward end to the center of the plug *D* is 12 inches. There is a back taper to this reamer of 0.001 inch per inch; consequently it is 0.004 inch smaller at *E* than at the forward end, or 0.002 inch on a side. The point *E*, then, will be 0.0013 inch lower than *F*, and theoretically will cut double this amount, *i.e.*, 0.0026 inch on the diameter.

Lest it be said that an almost impossible case has been used

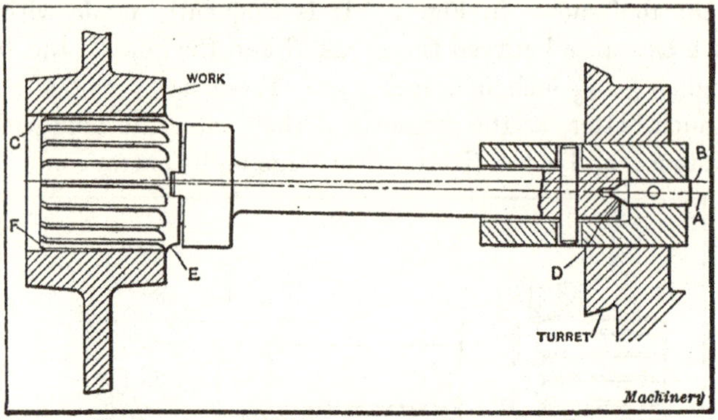

Fig. 4. Disadvantage of Holder shown in Fig. 3 when Turret does not index Properly

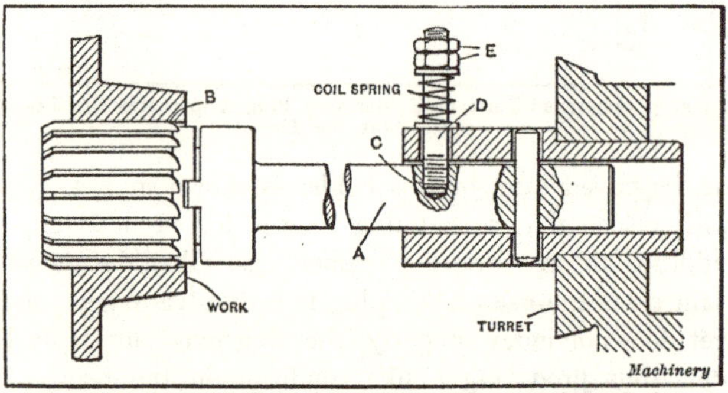

Fig. 5. Method of Balancing a Reamer Holder with a Spring

as an example, the author would say that about two years ago a condition even worse than this came under his observation. A turret lathe of a well-known make had been in use on cast-iron work for only about six months and yet the center of the turret hole was 0.014 inch lower than the center of the spindle, and it was impossible to produce a correctly sized hole upon it, until the spindle bearings had been scraped to correct alignment with the turret. After this had been done, no further trouble was experienced.

An instance of a method of support by means of a spring is shown in Fig. 5, the holder being of the same general type as the one shown in Fig. 2. The reamer in this case was $3\frac{1}{2}$ inches in diameter and the shank *A* rather long. In testing the equipment preparatory to shipment, it was found impossible to obtain a correctly sized hole at the end *B*, where the reamer began to cut. It was bell-mouthed to the extent of 0.003 to 0.005 inch, for about half an inch back, and the entire hole was in-

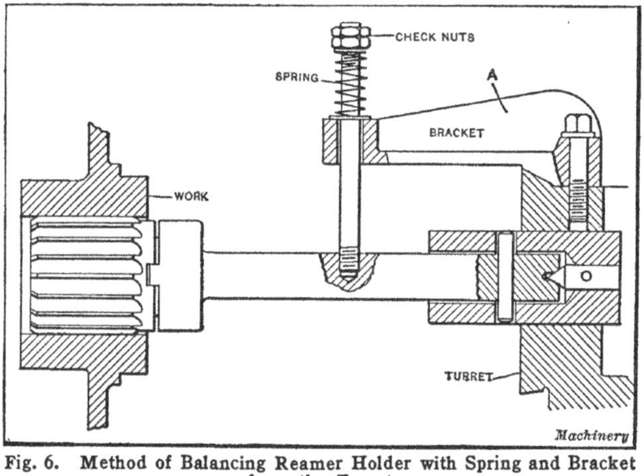

Fig. 6. Method of Balancing Reamer Holder with Spring and Bracket from the Turret

clined to be over size. The weight of the reamer being considerable and the shank rather long, caused a "drag" in entering the hole. The shank was drilled and tapped at *C* for a $\frac{1}{2}$-inch stud, and a $\frac{5}{8}$-inch clearance hole was provided in the holder directly above it at *D*, a stiff coil spring being placed in position as shown. The check-nuts *E* were then used to compress the spring until the weight of the reamer was properly balanced. This arrangement was found to correct the trouble and the hole obtained was sized properly. Fig. 6 represents a similar case where the reamer was still longer and heavier, but in this instance the supporting bracket *A* was fastened to the turret itself, instead of taking the thrust of the springs directly on the holder.

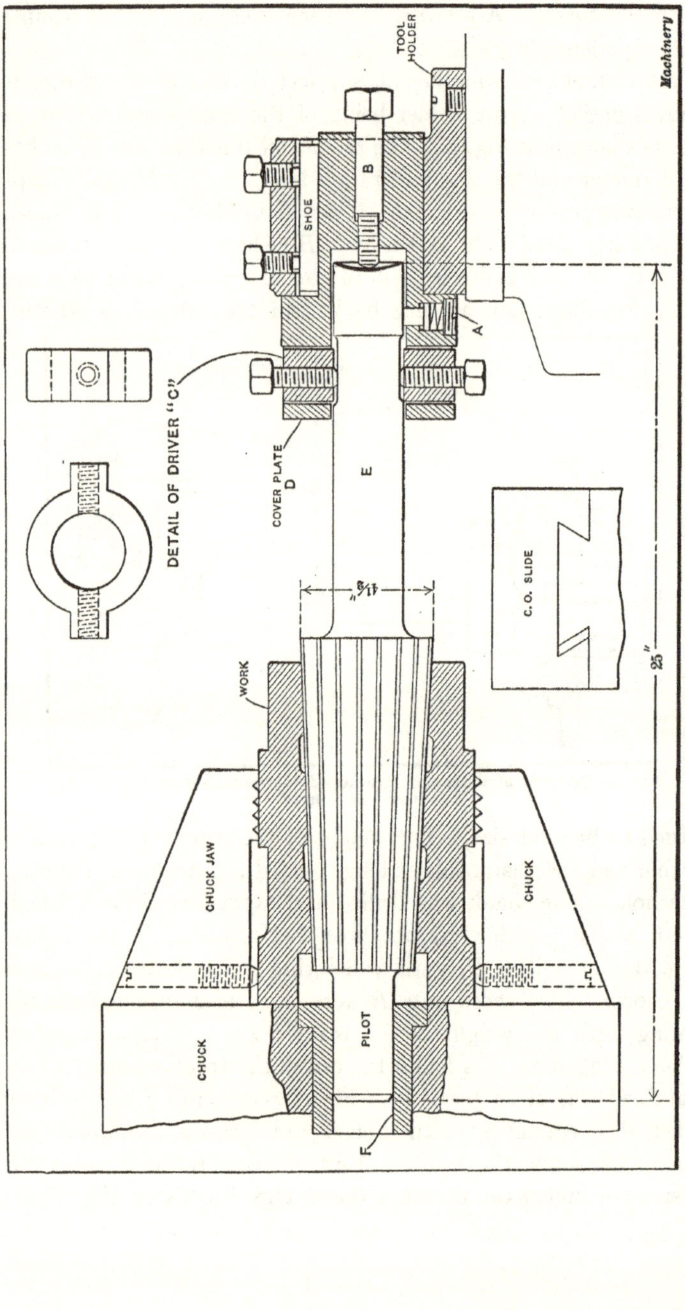

Fig. 7. Large Taper Reamer and Method of Balancing

Holder for Large Taper Reamer. — The holder shown in Fig. 7 was made for a very heavy piloted taper reamer of the dimensions indicated in the illustration. The cross-slide was used during the setting of the work, and it was necessary to provide a long shank for the reamer in order to avoid interference. This increased the weight to somewhat over 40 pounds, and it will be readily seen that some sort of support for the shank was essential to prevent a dragging action in the hole. For this purpose three spring shoes were provided, 120 degrees apart (one of which is shown at *A*). The pressure on the shoes

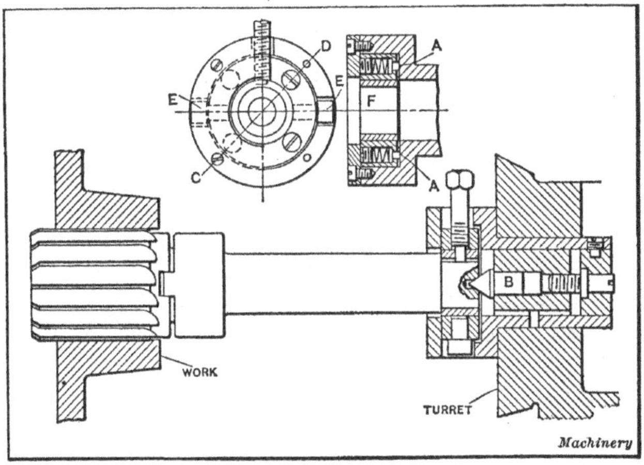

Fig. 8. Pratt & Whitney Reamer Holder equipped with Spring Shoes and Center Support

was controlled by the backing-up screws which assisted in keeping the shank in a central position. The thrust of the cut was taken by the screw *B*, and the driver *C* (a detail of which is shown in the upper portion of the drawing) took care of the torsional strain. The clearance hole in the cover plate *D* was only 0.004 inch larger than the shank *E*, thus permitting the end of the pilot to enter the bushing F without assistance. Had this clearance hole not been made in this way, it would have been a difficult matter to start the pilot properly, because of the weight which would have had to be lifted and started by hand. This holder gave very satisfactory results.

Pratt & Whitney Reamer Holder. — The Pratt & Whitney Co., Hartford, Conn., manufactures a holder of a somewhat similar construction (see Fig. 8) in which equalizing spring shoes are also employed, as shown in the upper portion of the illustration at *A*, the section being taken along the line *C–D*. An adjustable and renewable center support *B* is an additional refinement. The bushing *F* may be easily made to fit various sized shanks. The drive provided through the pins *E* is exceptionally powerful.

Holder for Small Reamers. — Fig. 9 represents a type of holder suitable for small reamers, which may be so designed

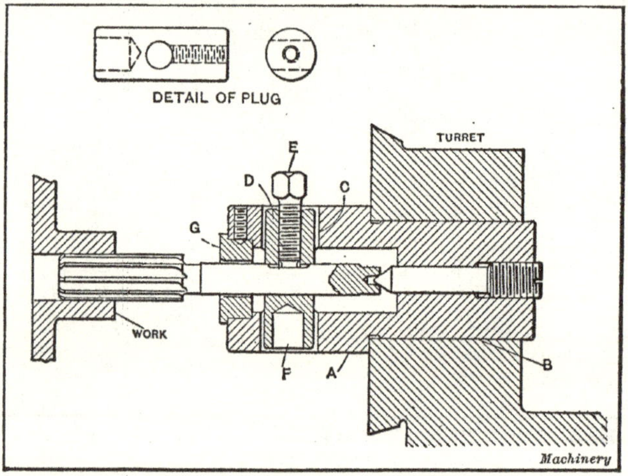

Fig. 9. Reamer Holder adapted for Use on Small Hand Screw Machines

that standard tools of this kind of various sizes may be easily used. The body *A* is made from square stock of suitable size turned down at *B* to a diameter corresponding with the hole in the turret. A piece of cold-rolled steel *D* is drilled and reamed to receive the reamer shank, and a set-screw provided at *E* to hold it in place. The drilled hole at *F* is for the purpose of removing unnecessary weight from the plug. It is obviously an easy matter to make up a set of plugs for different sizes of shanks and a series of bushings *G* to correspond. These bushings control the amount of float in the reamer, and, speaking generally, from 0.003 to 0.005 inch should be sufficient clearance between the bushing hole and the reamer shank. This type of holder is extremely useful on hand screw machines, and although it possesses a centering plug, it is less likely to cause inaccuracy, as its use is confined to the smaller machines where there is less wear and tear and consequently less chance for variations in alignment.

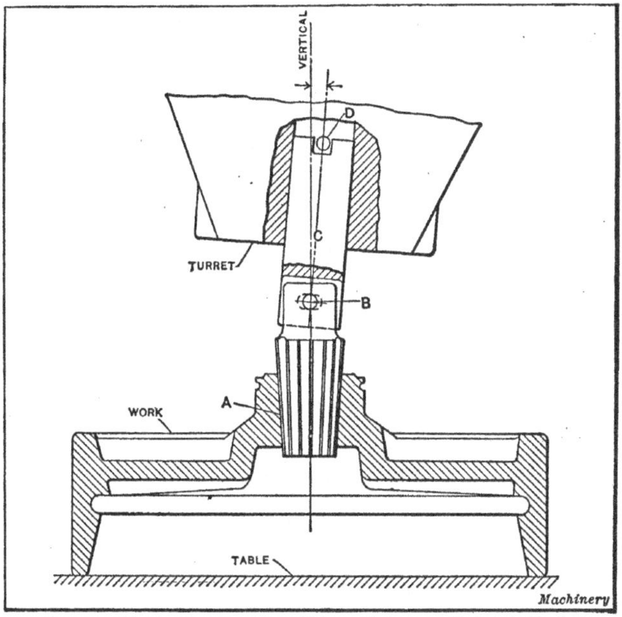

Fig. 10. Holder for reaming Taper Holes on Bullard Vertical Turret Lathe

Holders used on Vertical Turret Lathes. — Fig. 10 shows an entirely different type of holder which operates under quite different conditions. It is used on the Bullard vertical turret lathe when the swivel slide carrying the turret has been set over at an angle for the purpose of boring the taper hole *A* in the automobile flywheel shown in the illustration. The reamer hangs vertically from the removable pin *B*, which is free to float in the slotted shank *C*. This shank fits the turret hole and is prevented from twisting by the pin *D*. Obviously the position of the turret "observation stops" must be determined with some care, in order that the reamed hole may be of the correct angle and not distorted by crowding over and thus "cocking" the reamer. The correct position may be easily determined by setting the reamer snugly in place in a correctly sized hole, and then bringing the turret down until the pin *B* passes freely into its proper place. The necessary stops are then set for this location.

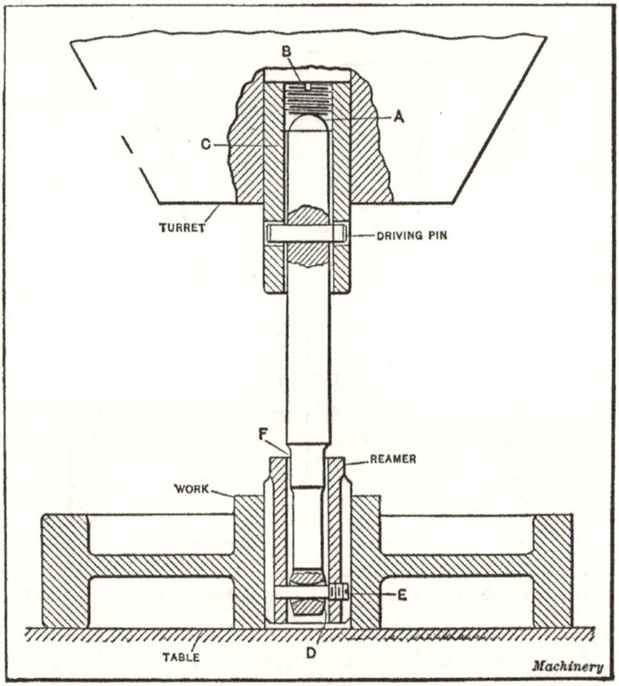

Fig. 11. Reamer Holder adapted for Use on either Horizontal or Vertical Machines

Another type of holder which is used in either a vertical turret lathe or in a horizontal machine is shown in Fig. 11. It has a ball-end shank *A* thrusting against the backing-up screw *B* in the end of the holder *C*, through which the driving pin passes. The forward end of the shank *A* is also ball-ended at *D*, and the screw-pin *E* passes through this portion, thus securing the reamer in position. It will be noted that the hole in the shank, through which this pin drives the tool, is bell-mouthed on each side of the center, and that there is also clearance at F between the shank and the reamer shell. This holder gives a free floating action in any direction, and comes

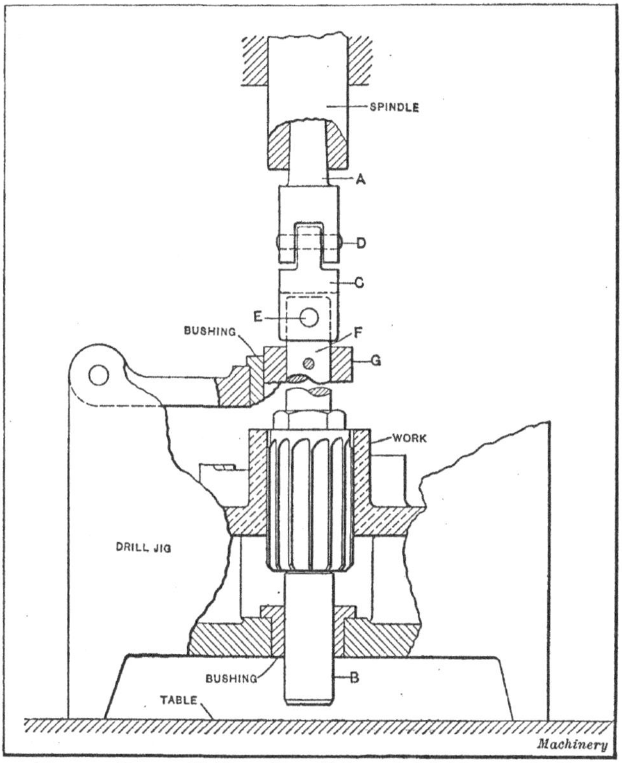

Fig. 12. Reamer Holder with a Universal Joint to provide the Floating Feature

nearer to the ideal type than any of those previously mentioned. It was manufactured at one time by the Bullard Machine Tool Co., Bridgeport, Conn.

Reamer Holder with Universal Joint. — Fig. 12 represents a holder having a universal joint, used in a drill jig of the type shown in the illustration. The shank *A* is made to a standard taper corresponding to that of the drill press spindle, and the piloted-end *B* is guided by the bushing in the jig. One end of the intermediate link *C* is milled flat and is hung loosely on the pin *D* through the slotted shank. The other end of this link is slotted at right angles to the first-mentioned flat and serves to drive the reamer through the connecting pin *E*, which passes through its shank F. A hardened and ground collar *G* is also pinned to the shank and is guided by the upper bushing in the jig. Various adaptations of this universal joint are seen in connection with fixtures used on horizontal boring machines, and are often made with some device for rapidly interchanging bars, reamers, etc., in place of the pin *E*.

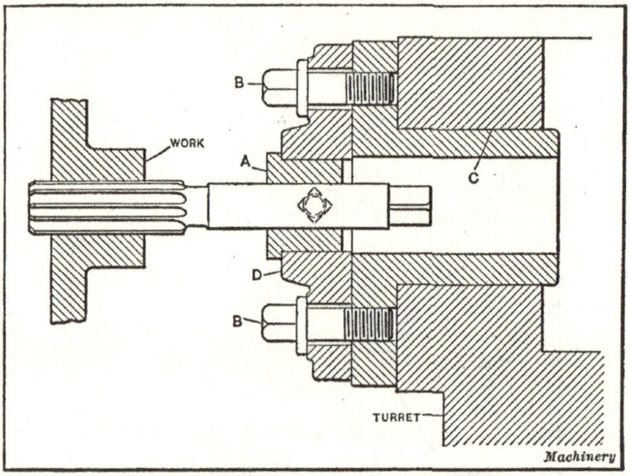

Fig. 13. Reamer Holder with Means of Adjusting it to Spindle Alignment

There is another holder (see Fig. 13) which is sometimes erroneously referred to as a "floating holder," but it has no "float" whatever, although it does have both vertical and horizontal adjustment. It is extremely useful in automatic screw machine work. The adjustable cap *D* may be quickly set to correct any inaccuracies of alignment between the turret and spindle by loosening the two screws *B*. The holes in the cap are made $\frac{1}{16}$ inch over size to allow for this, and the bushing *A* is fitted to the reamer shank. The body *C* of the holder fits the hole in the turret and is secured by some approved method.

CHAPTER V

ARBORS FOR TURNING, BORING AND GRINDING

Methods of Holding Work for Machining. — Cylindrical work which cannot be completely machined in one setting and which requires concentricity of the various surfaces obviously makes necessary some method of holding it for the second operation which utilizes a previously machined surface for securing the proper location. When this surface is external, the use of soft jaws, a step-chuck or collet jaws is feasible, but when an internal surface is the locating point the most efficient method is conceded to be some form of arbor. This arbor may be either a plain stud made to fit the hole in question, or it may be so designed as to be susceptible of a certain amount of expansion and contraction in order to take care of slight variations in the finished hole. The degree of accuracy required in the finished product determines the form of arbor which should be used. If a variation of 0.002 to 0.003 inch in concentricity is permissible, a plain arbor with some method of driving the work will answer the purpose very well. When very accurate work is required, however, greater care must be used in the design, and the expanding type of arbor is commonly used.

Important Points in the Design of Arbors. — The fundamental features which tend to make an arbor thoroughly efficient are as follows: Expansion must be uniform along the entire periphery; release must be quick and easy; ample driving facilities must be provided; clamping the work must be effected without chance of distortion. As an additional refinement, provision may be made for truing up the arbor so that it will run accurately with the center line of the spindle.

Lathe Arbors. — Let us consider the arbors designed for use in the engine lathe, adapted to be held between centers and driven by means of a dog on one end. The arbor shown in the upper part of Fig. 1 is the simplest of all those which have a split sleeve or bushing capable of expansion or contraction. The mandrel *D* is slightly tapered and is flattened on each end to receive the dog for driving. The sleeve *C* is correspondingly tapered and is drilled entirely through at *A*, after which it is saw-cut at *B* to allow for expansion. This arbor is the poorest of all the expanding types in that the expansion is not uniform, being in two directions only, and it cannot be depended upon to give results which are absolutely accurate.

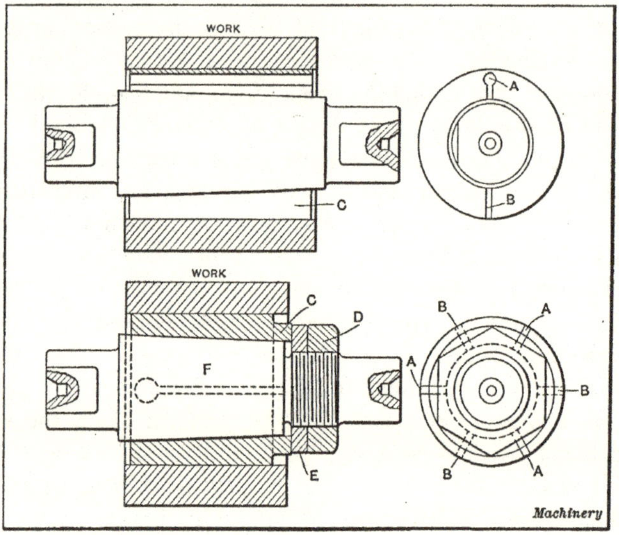

Fig. 1. Two Types of Expansion Arbors with Split Sleeves

A much better arbor is shown in the lower part of the same illustration. It will be noted that the mandrel *F*, in addition to being tapered, is threaded on one end to receive the hexagon nut *D*. This does away with the necessity of using the arbor press to expand the sleeve. The collar *E* is interposed between the nut and the split bushing. This bushing *C* is saw-cut at *A* from one end and at *B* from the other, thus allowing a uniform expansion along its entire periphery. In this connection, it is well to note that the ends of the saw-cuts should be left tied together until after the sleeve has been hardened and ground; they can then be cut apart readily with a thin emery wheel. An arbor of this kind is mechanically correct and, if carefully made, should give results which leave nothing to be desired as far as accuracy is concerned.

Fig. 2 shows an arbor of a very different type, which might be called a solid expanding arbor. Three holes *B* are drilled 120 degrees apart and the saw-cuts *A* are milled as shown. The special screw *E* is tapered at *C* and threaded at *D* in the body of the arbor. The end of this screw is squared and con-

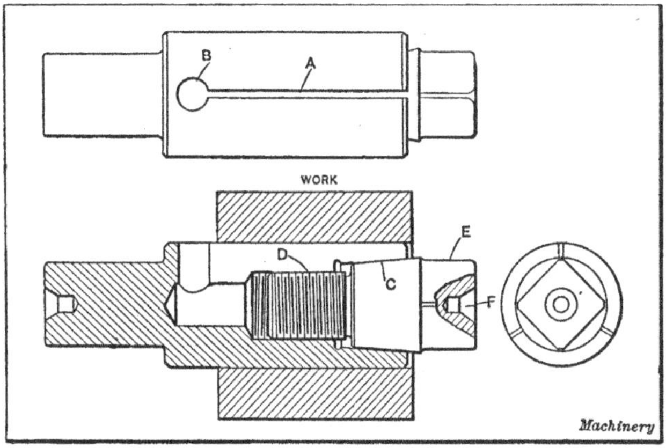

Fig. 2. Arbor expanded by Internal Taper Plug

tains the center F. When made as shown there is nothing in this arbor to commend it. In the first place, the expansion takes place at one end only and is not at all uniform, and, in the second place, the center F in the end of the screw cannot be depended upon to remain true for any length of time, even assuming that it may have been made reasonably true to start with, which, in itself, is a difficult machining proposition.

The arbor shown in Fig. 3 was at one time manufactured commercially by G. E. Le Count, South Norwalk, Conn., but the author is unable to state whether it is on the market at the present time or not. It consists of the body *A* in which are milled the tapered slots *B*. The shoes *C* (also shown in detail in the upper part of the illustration) have a narrow rib running along each side and this rib engages with the grooves in the sides of the slots *B*, thus preventing the shoes from falling out. The collar *D* controls the action of the shoes and is ground to a sliding fit on the cylindrical portion *E* of the arbor. It may be noted that the shoes have two shoulders, thus increasing the range of the arbor. By providing shoes of various diameters the range can be increased considerably.

W. H. Nicholson & Co., Wilkesbarre, Pa., manufacture the expanding arbor shown in Fig. 9 in various sizes and to suit various conditions. The body *A* is made of tool steel, hardened and ground to a cylindrical form. The centers are

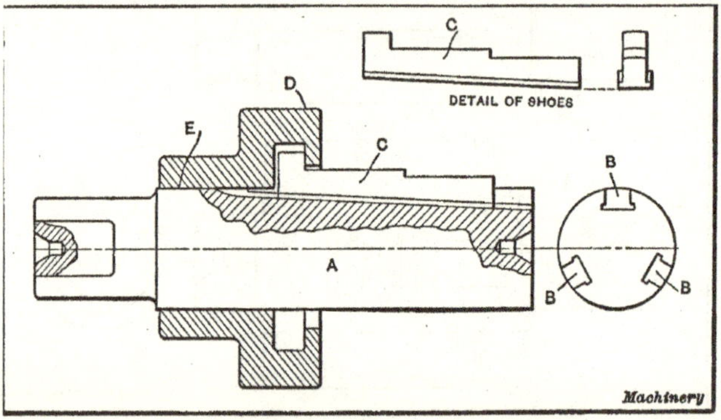

Fig. 3. Expanding Arbor of Sliding-shoe Type

exceptionally large and are carefully rounded and lead-lapped after hardening. There are four slots in the body (shown in the section *A–B*), and these slots are relieved at each corner to prevent any interference by dirt. After hardening, the slots are also ground to insure truth. The jaws *C* (also shown in detail) are made of special steel and carefully ground to the same taper as the slots. After assembling, they are also ground radially on their own arbor. The sleeve *D* acts as a retainer for the jaws and is a running fit on the cylindrical portion *E* of the arbor. Four slots are cut through the sleeve and the jaws are held in position by them. These arbors are too well known to need further comment, as they are in general use throughout the country.

Turret Lathe Arbors. — We will now go a step further and take up the type of arbors adapted for use in the horizontal turret lathe. It is well to bear in mind that arbors of this sort should be so designed that the work may be easily and quickly put on and taken off without the assistance of anything more than a wrench or spanner. Every precaution must also be

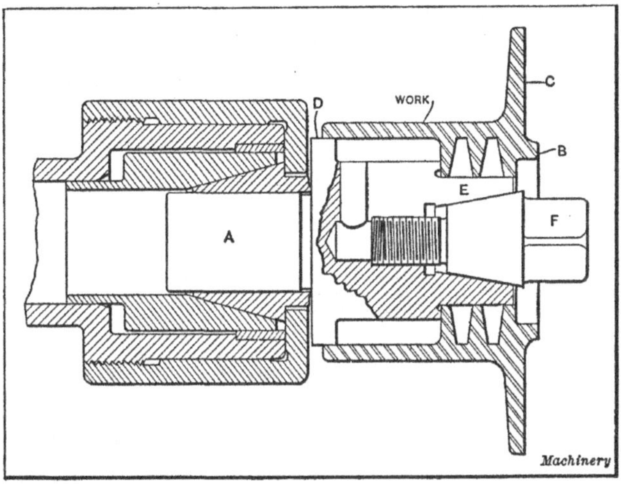

Fig. 4. Arbor held in Collet and expanded by Internal Taper Plug

used in clamping and driving the work, so that no chance for distortion is possible.

The arbor shown in Fig. 4 is somewhat similar in construction to that in Fig. 2, except that it is adapted to be held in collet jaws instead of on centers, as in the former instance. This arbor gave satisfactory results on the work for which it was used, the surfaces *B* and *C* being faced within the required limits of accuracy. The work was a push fit on the cylindrical portion *D*, the expansion taking place at *E*, controlled by the tapered screw F. In this case, the nature of the work permitted a slight margin of error and the expansion was only necessary to prevent chatter and act as a driver. The shank *A* was held in the collet jaws.

The arbor shown in Fig. 5 was made for the transmission gear which is shown in position. After the taper hole had been "chucked" in the work (which was done in a previous setting), the keyway *A* was cut for assistance in driving. The arbor body *B* is of cast iron, ground to fit the spindle at *C* and *D*. The stem *G* is of steel, hardened and ground to fit the body, into which it is keyed to resist the torsion of the cut. It is held in position and drawn back by the nut and collar *L* and *M*. The forward end is ground to the correct taper *E*, and the key is inserted at *A*. The portion *F* is threaded with a six-pitch Acme thread, right-hand, and the nut *H* is used to remove the piece after the work is finished, a piece of drill rod being used in the spanner hole *K* to turn the nut. The screw *N* prevents

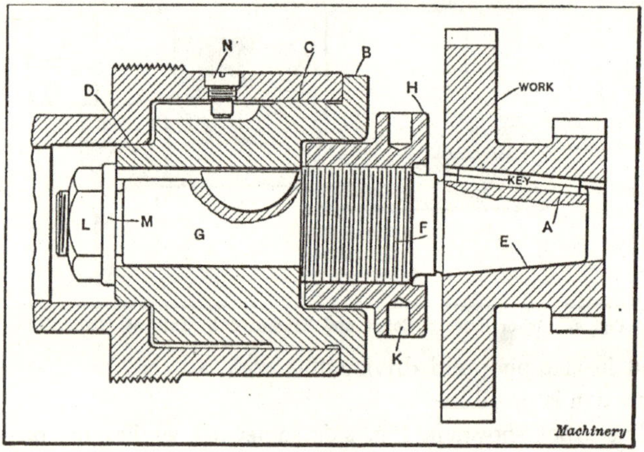

Fig. 5. Taper Arbor mounted in Spindle and equipped with Nut for removing Work

the body from turning in the spindle. This arbor has given very satisfactory results.

A somewhat extraordinary condition is shown in Fig. 6, which illustrates a steel automobile hub and arbor. In this case, the bearing seats at *A* and *B* were required to be absolutely concentric. In order to assist in machining, the portion *C* was bored to size in the first setting, although no finish was required at this point. The body of the arbor *D* is of tool steel, hardened and ground at all important points. The small end is slotted at three places as shown at F and is spring tempered at this end. The operating rod *K* has a very free thread at *H* and is ground to a snug running fit in the cylindrical portion *G* to insure true running, regardless of the condition of the threaded part. All the tools used on surface *B* of the work were piloted by the stem or extension provided on rod *K* in the direction towards the right, thus securing absolute truth and concentricity of the ends *A* and *B*.

There are several important points to be noted in the construction of this arbor. First, the method of obtaining a true running stem or extension of rod *K* by means of the long cylin-

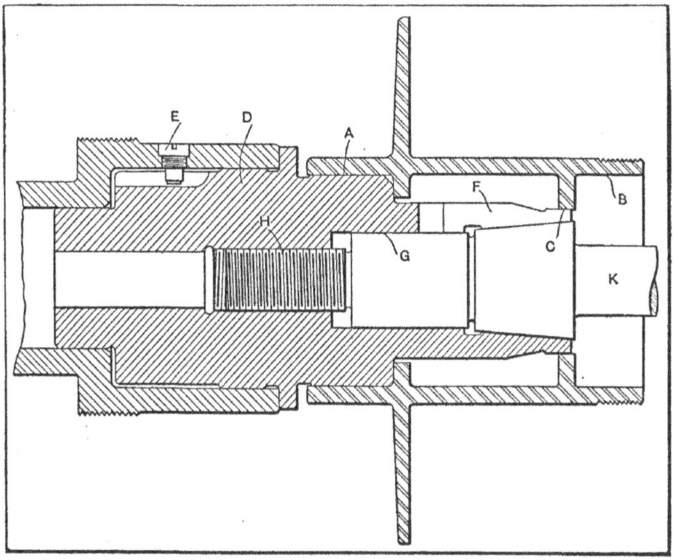

Fig. 6. Arbor with Split Expanding End and Pilot for Steadying Tools

drical bearing at *G*; second, the use of the stem as a pilot for tools, thereby obtaining concentricity in the two ends of the work *A* and *B*; third, the positive location of one end at *A*, while using an expansion principle at the other end to insure rigidity and freedom from chatter. This arbor was very satisfactory, the two ends being within the extremely narrow limits of concentricity required.

An entirely different type is illustrated in Fig. 7. This is used for two different sizes of bronze bearing retainers, the use of adapters making this feasible. In the construction of this arbor, the body *A* is screwed directly onto the spindle nose, bringing up snugly against the end of the spindle at *B*. The body itself is of steel and is tapped out at *C* to receive the operating screw *E*. As in Fig. 6, the thread is a free fit, while the cylindrical portion *D* is ground to a snug running fit to insure concentricity. The bushing *F* is saw-cut in six places, three cuts from one end running nearly through, and the other three in like manner, in order to allow uniform expansion of the bushing. Both the bushing and the operating screw are tapered correspondingly at *G*. The adapter *H* slips onto the body of the

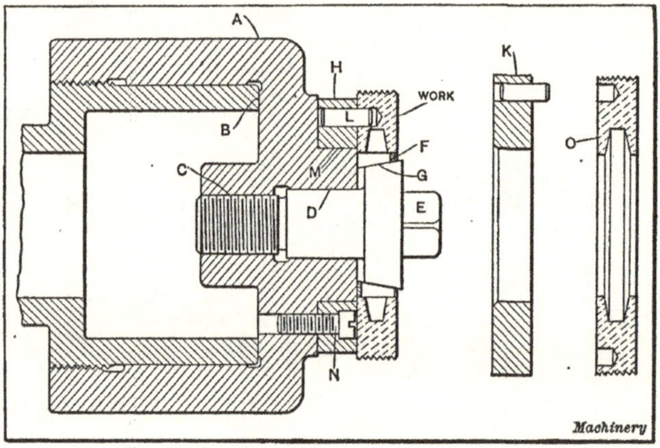

Fig. 7. Arbor having Taper Plug which expands Split Bushing

arbor and is located from shoulder *M* and secured in place by three screws *N*. A pin driver *L* in the adapter relieves the bushing of excessive strain. The larger retainer *O* (shown in detail) is also handled on this same arbor by using the adapter *K*. The results obtained with this arbor were perfectly satisfactory.

In Fig. 8 the principles of expansion and contraction are both used in handling steel rifle part *A*. The permissible limits of error on this work were very close, so that extremely careful workmanship was necessary. This operation was the final one on the piece, after it had been machined all over, leaving

0.015 inch at *C* for truing to insure absolute concentricity between surfaces *B* and *C*.

The machine to which this fixture was applied was equipped with collet mechanism, part of which was used in the operation. The body of the fixture *E* is of cast iron and is screwed onto the spindle nose, being secured against turning by the teat-screw F. The sides of the fixture were left open to enable the operator to reach in and grasp the work, in order to guide it onto the locating bushing *B*. The operating rod *G* was milled at an angle on the forward end *H*, in order to force the pin *K* outward and thus insure rigidity at this point. The collet operating sleeve *M* is secured to the rod by pin *N*, so that the

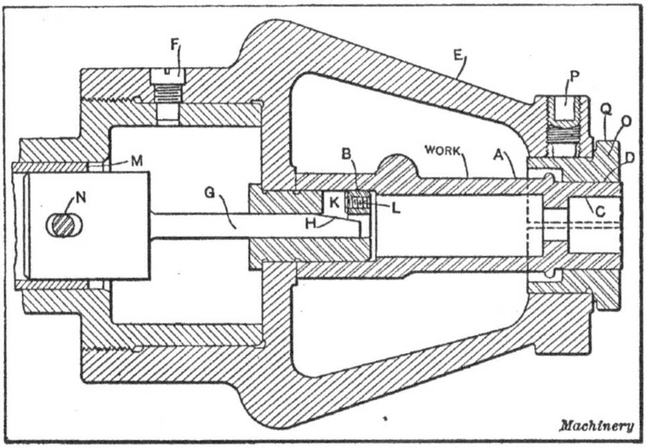

Fig. 8. A Fixture which holds Work by Expansion and Contraction

collet closing mechanism can be used to operate the rod. At the forward end of the fixture, the split bushing *O* is used to center the work which is gripped on the finished cylindrical surface *D*. The bushing is knurled at *Q* and is contracted by the action of the hollow set-screw *P*. All important surfaces were ground to an accurate fit and parts subject to wear were hardened. No trouble was experienced with this fixture and the work was machined within the limits of accuracy required.

The steel pinion blank shown at *A* in Fig. 10 has been previously faced at *B* and the taper hole carefully bored, leaving the remainder of the work to be accomplished at the setting

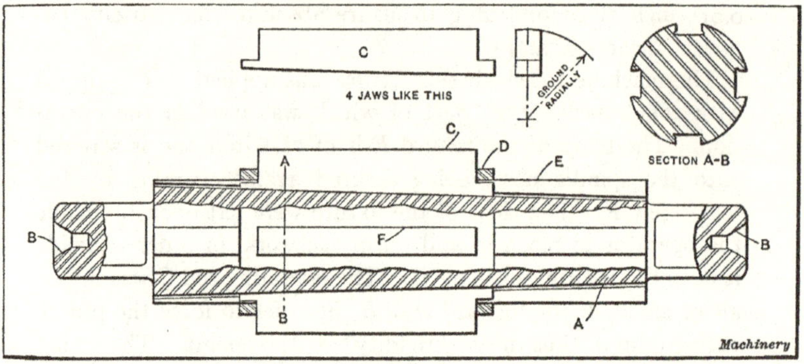

Fig. 9. Expanding Arbor with Sliding Shoes retained in Slotted Sleeve

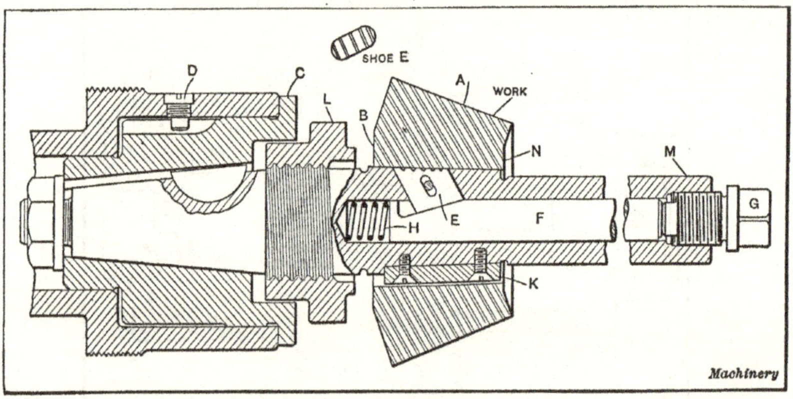

Fig. 10. Arbor having Adjustable Shoe E bearing against Taper Bore of Pinion Blank

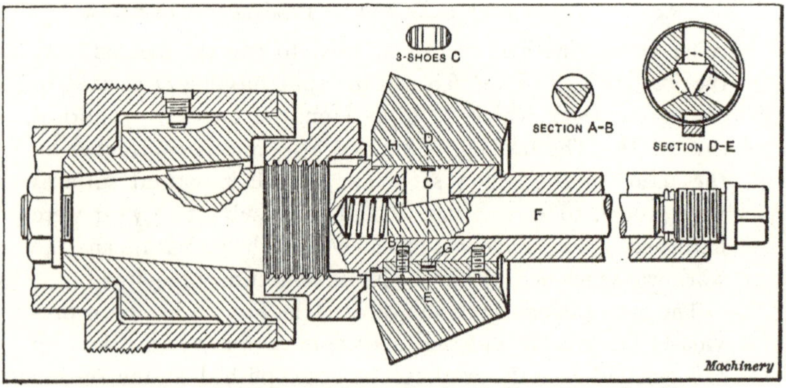

Fig. 11. Arbor for Pinion Blanks equipped with Three Expanding Shoes

shown. The body *C*, in this instance, is of cast iron and is held in position by the teat-screw *D*. The arbor is of tool steel, carefully hardened and ground. The shoe *E* is shaped as shown in the detail above. The operating rod *F* is forced inward by screw *G*, and its release is effected by spring *H* which bears against its inner end. It should be noted that the action of shoe *E* is both outward and backward; therefore it has a tendency to force the work back onto the tapered portion. Obviously key *K* acts as a driver. In order to avoid any chance of springing the arbor out of true a small, special wrench is used for turning screw *G*, so that too much pressure cannot be applied. The nut *L* is threaded on the arbor with a coarse-pitch Acme thread and is of hexagon shape at the forward end. This nut is used to start the work off the arbor when the piece is finished. The stem of arbor *M* enters a bushing in the turret and acts as a support while the beveled surface of the work is being turned. This stem is also used as a pilot for the face mills which form the end of the pinion at *N*. This arbor while used for producing work of the best quality was somewhat fragile and required careful handling.

The pinion blank shown in Fig. 11 has a straight hole instead of a taper one, and the arbor for holding it, while somewhat similar in construction to that shown in Fig. 10, differs as regards a number of points. There are three shoes *C*, 120 degrees apart, controlled, as to their outward movement, by the operating rod *F*. These shoes are retained in their positions by the thin circular spring *G*. The shoulder *H* on the arbor acts as a positive longitudinal stop for the work. The various sectional views give a good idea of the construction. This arbor is also of tool steel and all important surfaces are hardened and ground. It gave very satisfactory results, but the rather delicate construction necessitated careful handling.

Arbors for Vertical Boring Mill and Vertical Turret Lathe Work. — We now come to a class of work of larger size, which can be more conveniently handled in the vertical boring mill or the vertical turret lathe. Arbors for comparatively large work frequently develop into combination locating and holding devices, so that they are more nearly related to locating fixtures in the truest sense of the word. It is well to remember that in all fixtures of large size some efficient means of driving the work must be provided, for the thrust of the tool, incident to the cutting action, is much greater on work of large

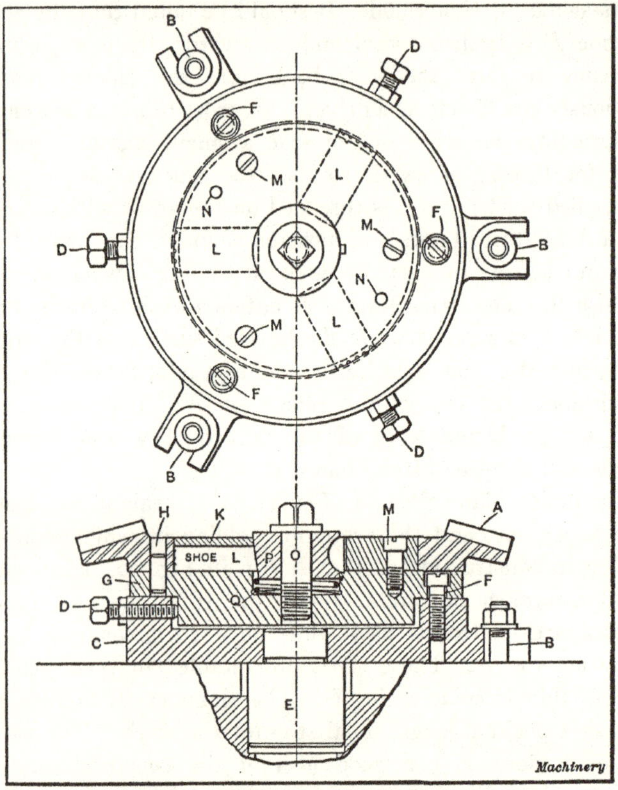

Fig. 12. Vertical Boring Mill Fixture for Holding Bevel Gear — The Three Shoes L are forced outward by Plug in Center

diameter; furthermore, the amount of stock to be removed is usually considerably more than on smaller work. The fixtures themselves should also be of exceptional strength and rigidly secured to the table to prevent movement or breakage.

The large automobile bevel driving gear shown in Fig. 12 is of alloy steel, and has been previously bored and faced on the rear side; the screw holes were also drilled in a jig before placing the gear on the fixture shown. This fixture was rather expensive, being made entirely of steel (except the base, which is of cast iron), and all working parts were hardened and ground or lapped to a perfect fit. The base is located in the center of the table by means of the locating stud *E*, and is securely fastened down by the three T-bolts *B* which enter T-slots in the table. The adjustable part *G* is held onto the base by the three screws F. It will be noted that the screw holes have a certain amount of clearance over the body of the screw to permit adjustments to be made. The screws and check-nuts at *D* are for the purpose of conveniently adjusting the fixture. A pin-driver *H* engages one of the jig-drilled holes in the work. The upper plate *K* is square slotted at three points *L* to receive the shoes *L*. The screws *M* and the dowels *N* hold this plate in its proper position. The collar-head screw *O* forces down the plunger *P* on which three angular flat spots are milled. These angular surfaces control the action of the shoes *L* and force them out, uniformly, against the work, thus centering it. The spring *Q* simply aids in releasing the shoes.

This fixture is an exceptionally good one, for in its construction every care is taken to insure a true-running arbor and one which can readily be indicated for truth and brought into perfect concentricity by means of the adjusting screws. Its action was satisfactory in every respect.

Fig. 13 shows an automobile flywheel which has a finished taper hole and has been turned, bored and faced in a previous setting. It was essential that the surface *A* should be concentric with the taper hole. As it was practically impossible to machine the face of the flywheel *B* and the taper hole *C* so that they would always come in exactly the same relation to each other, it was necessary to make the taper plug adjustable in a vertical plane. The base of this fixture is of cast iron and is located centrally by means of plug *D* which accurately fits into the hole in the table. The base is clamped in position by three T-bolts *E* (see plan view) engaging the table T-slots. Three clamps F are used to clamp the work down on the annular rim *G* of the fixture. The plug *D* not only locates the fixture base, but extends above the latter and is threaded at *H*, while above this portion it is cylindrical and is carefully ground to a running fit in the taper bushing *K*. The threaded portion mentioned is a very free fit, so as to permit the cylin-

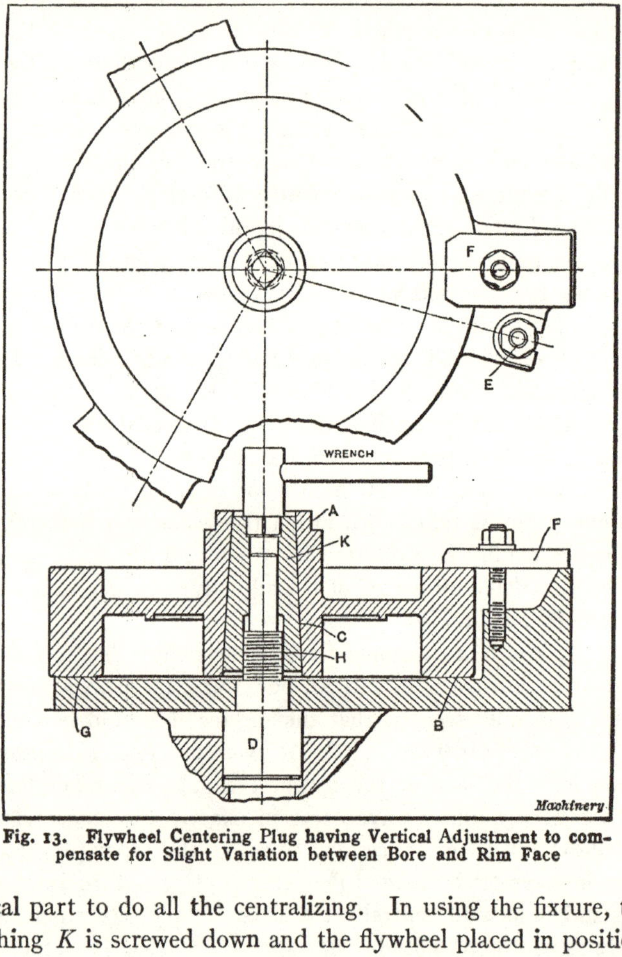

Fig. 13. Flywheel Centering Plug having Vertical Adjustment to compensate for Slight Variation between Bore and Rim Face

drical part to do all the centralizing. In using the fixture, the bushing *K* is screwed down and the flywheel placed in position, after which, by the aid of the wrench, the bushing is raised until it bears in the taper hole. After this, the clamps *F* are swung around and tightened. This is a simple fixture, rather inexpensive, and one which was thoroughly dependable.

Fig. 14 shows a cast-iron double-bevel gear used on harvesting machinery, the gear rings *A* and *B* having cast teeth. These

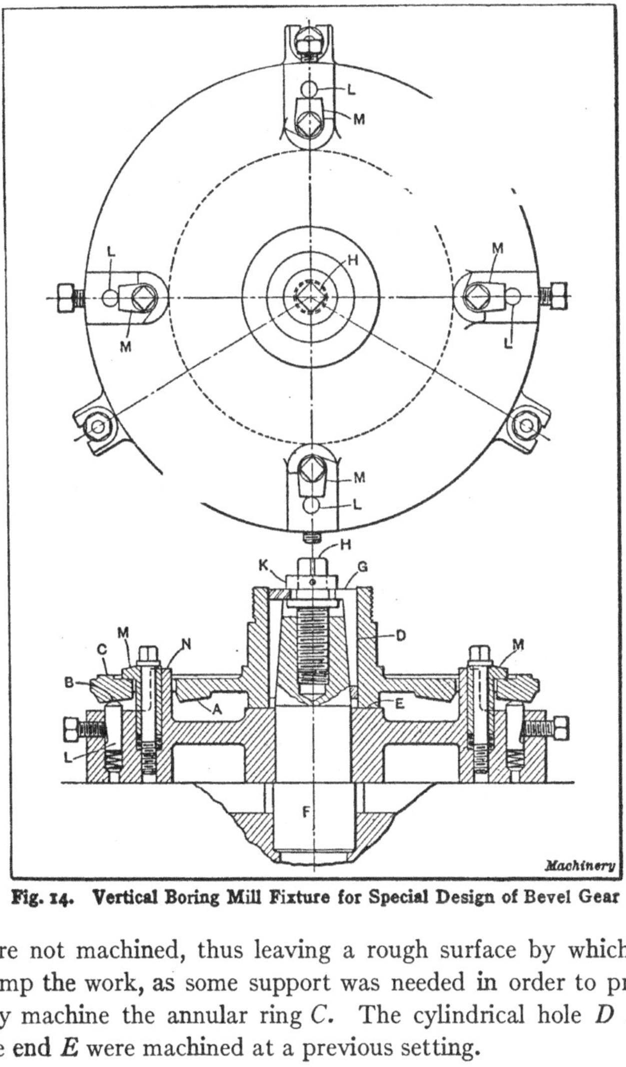

Fig. 14. Vertical Boring Mill Fixture for Special Design of Bevel Gear

were not machined, thus leaving a rough surface by which to clamp the work, as some support was needed in order to properly machine the annular ring *C*. The cylindrical hole *D* and the end *E* were machined at a previous setting.

The cast-iron base of the fixture is centered by the stud F which fits the center hole in the table. This stud extends up through the fixture and is tapered at its upper end to receive the split bushing *G*. This bushing is saw-cut in six places — three from each end — and is shouldered at its upper end so that the vertical movement can be controlled by the operating sorew *H*. The collar *K* was pinned in place after the bushing was slipped over the screw. It will be noted that the vertical movement of the bushing is entirely mechanical, no springs being used to effect its release, as in previous instances. The positive locating point of the fixture is at *E*, but as it was necessary to have some support at *B* the four spring pins *L* were used; these bear against the rough surface of the casting and are prevented from being pushed down by the screws shown. The flat spot against which the screws bear is milled back at an angle of ten degrees. The rim of the gear has four cored holes and hook-bolts were necessary for holding and driving. These are shown at *M* in the illustration. In this connection, it is well to note that these hook-bolts are well backed up by a portion of the fixture *N*, for a hook-bolt which is not backed up in some way is worse than useless. This fixture was capable of rapid manipulation and the results obtained by its use were within the necessary limits of accuracy.

An Expanding Arbor for the Vertical Milling Machine. — In one of the large automobile factories, considerable trouble was experienced in the manufacture of eccentric piston rings by the breaking of the rings as they were being cut off on an automatic machine. A new method was therefore devised by which the "ring pots" were bored and turned eccentric and then taken to a vertical milling machine where they were placed on the arbor shown in Fig. 15. The fixture of which the arbor forms a part is located in the center of a circular milling table by the stud *B*, and is secured to the table by means of the three screws *A* in the T-slots. The stud is tapered at its upper end to receive the split bushing *D*, and is secured by the pointed screw *C* and prevented from turning by the key *E*. The bushing was saw-cut in six places to permit expansion, and was also counterbored in three places at its lower end to make a pocket for the coil springs F. These springs tend to make the releasing of the split bushing easy after the work has been done. The collar *G* bears on the upper portion of the split bushing *D* and is operated by the screw *H* which is threaded into the body of the arbor.

A special arbor *K* in the spindle of the vertical milling machine was arranged with a gang of saw cutters properly spaced for

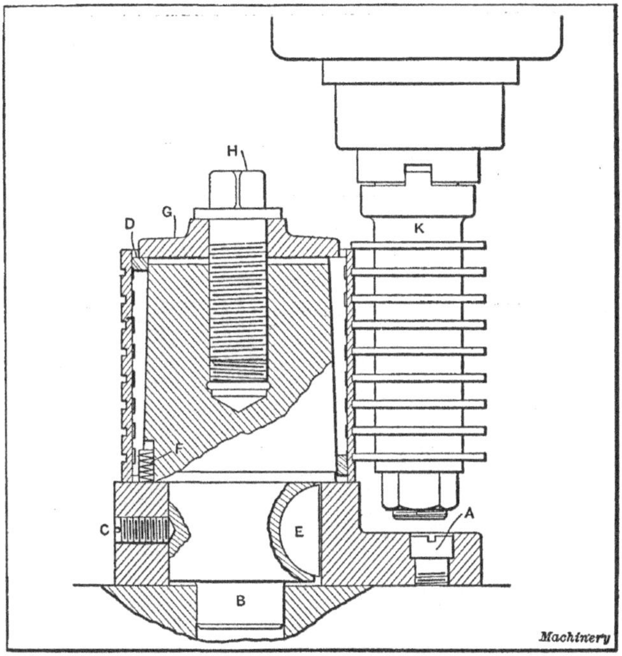

Fig. 15. Expanding Arbor for Holding Casting while Gang-sawing Packing Rings

the correct width of ring. As the table is revolved by power feed, the gang of cutters produce a set of nine clean and unbroken rings. It may be noted that the split bushing is relieved on its periphery at the points where the cutters pass through the work, in order to avoid dulling the cutting edges on the hardened surface. This fixture was made very carefully and proved very satisfactory.

Arbors for Threaded Work. — There are numerous instances in the manufacture of parts for interchangeable work, where a certain portion of the particular piece in question must be absolutely true and concentric with a thread which has been previously cut upon the other end. Sometimes an external thread has been chased or possibly cut with a die-head, while in other cases an internal part may have required the use of a tap to cut the thread. Occasionally both ends are threaded

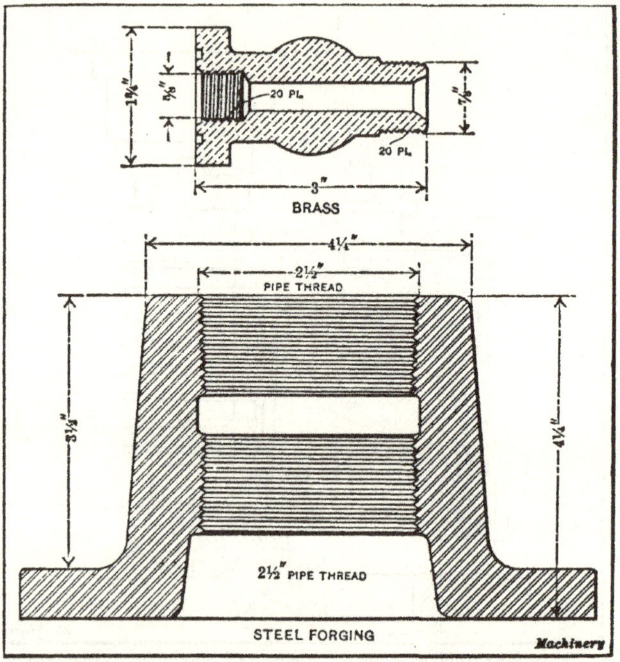

Fig. 16. Examples of Work handled on Knock-off Arbors

and must be concentric. Taper or pipe threads are found at times on a variety of work, and they also may have to conform to the same conditions. On some pieces the end adjacent to the threaded portion is left unfinished, which increases the difficulty in handling for any subsequent operations which may be necessary in the production of the work.

The machine tools available for the work being of various

types, and the work itself of widely different forms, obviously bring about the design of holding devices to meet the particular requirements of each condition. The horizontal turret lathe is probably more frequently used for this class of work than any of the other machines. The engine lathe is also frequently called upon in an emergency, while the vertical boring mill and the vertical turret lathe are used somewhat less frequently.

The work may range in size from a piece like the small brass pressure valve body shown in the upper portion of Fig. 16, up to the large steel boiler nozzle in the lower part of the same illustration. Larger pieces than this, and possibly still more difficult to handle, may be met with occasionally, but the fundamental points will be the same and will require the same methods of machining.

Elementary Design of Threaded or Knock-off Arbors. — Let us take the simplest condition possible, for example, the threaded collar shown at *A* in the upper part of Fig. 17, in which it is only necessary to face the end *B* so that it will be square with the thread. A plain arbor, such as that shown, with a shoulder against which the collar may be screwed, seems to meet the requirements of this case. But after the end has been faced, it is highly desirable to be able to remove the piece and put on another, and here lies the difficulty, for the wedging action at *C*, caused by the slight twisting of the work under the strain of the cut, has tightened the collar to such an extent that the assistance of a pipe-wrench (or some other method which may suggest itself) is needed before the work can be started. After this has been done, it will be found that the outer surface is considerably injured.

The arbor shown in the central illustration, Fig. 17, is a refinement of the first, in that the check-nuts *D* and *E* permit the easy removal of the piece after the work has been accomplished. No longitudinal stop for these nuts is provided, however, and therefore the length F is susceptible to variations, unless each piece is carefully measured to obtain the correct length while facing.

The lower arbor, shown in Fig. 17, overcomes the faults found in the two others and for lathe work it answers the purpose for which it is intended. The work *G* comes to a positive longitudinal stop against the face of the collar at *H*, and the collar itself has a fixed location in the shoulder *J*. A left-hand thread

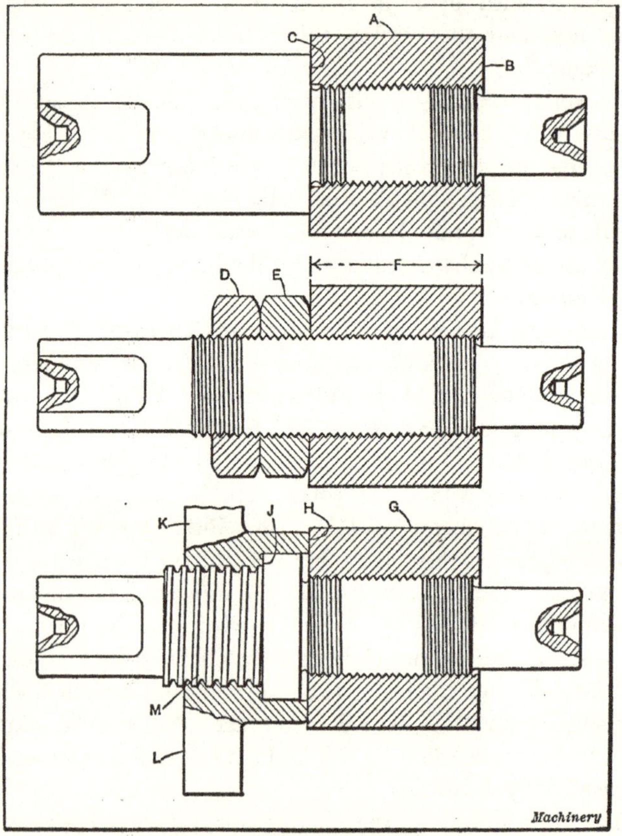

Fig. 17. Three Types of Arbor for Machining the Piece A

is provided on the arbor at *M*, and this effectually resists any tendency to unscrew, as the action of the tool upon the work is in the right direction to force it into this position. It should be noted that the two lugs *K* and *L* are integral with the collar, and forged as shown. After the machining operations have been completed, the work is released by a sharp blow with a babbitt hammer or a billet of wood on either of the two lugs.

In designing an arbor of this type, remember that the left-hand thread for the knock-off portion should be of very much coarser pitch than that of the work itself, because the wedging action is much less in the coarser pitch thread, and consequently the releasing may be more readily accomplished. A ratio of 1 to 2 is usually sufficient. For example, if the work is threaded 20-pitch right hand, it is advisable to thread the knock-off 10-pitch left hand or even coarser if desired.

The following are three important points to be observed in the design of arbors of this type:

First: Positive longitudinal location of the work.

Second: Means for minimizing wedging action and at the same time allowing quick and easy release.

Third: Construction of the arbor in such a way that no chance for springing out of truth is possible.

Knock-off Arbors for Small Work. — A small piece requiring a knock-off arbor is the brass valve body shown in the upper portion of Fig. 16. In this case it was necessary that the threaded portions should be concentric and true with each other, and it was not practical to make the piece from the bar complete in one setting. The arbor, shown in Fig. 18, was applied to the spindle of a small turret lathe, and a Geometric die-head, mounted on the turret, was used to cut the thread. The construction of the arbor was as follows: A special nose-piece *A* was mounted on the end of the spindle *B*, and a retaining screw *C* prevented it from unscrewing. The screw arbor *D* was keyed in position and drawn back firmly against the nose-piece by the nut shown.

The knock-off portion, in this instance, was (for assembling purposes) made in two pieces *E* and *F*, these being held together by three screws 120 degrees apart. The threaded portion of the work, shown at *H*, was 20-pitch right hand, while the knock-off, at *G*, was made 8-pitch left hand in order to minimize the wedging action and make releasing easier. In using this arbor, the knock-off was brought up against the shoulder *J* which constitutes the positive stop, and the work screwed on at *H*, the end bringing up firmly on the ring at *K*. The shoulder *L* and thread *M* were then machined in their correct relation, by turning and threading tools located on the turret. To release the work after the machining had been accomplished, a

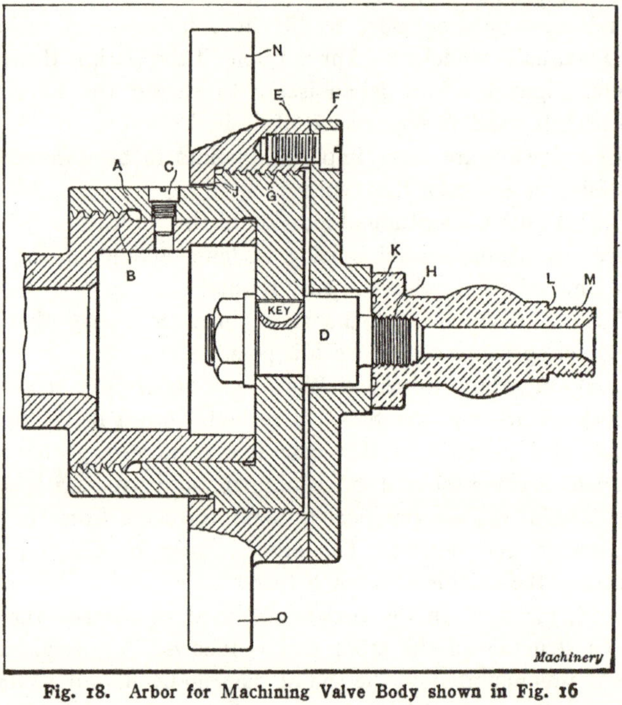

Fig. 18. Arbor for Machining Valve Body shown in Fig. 16

sharp blow with the babbitt hammer on either of the two lugs *N* or *O* was all that was necessary. After this the work could be readily unscrewed by hand.

Fig. 19 shows a knock-off arbor which is not used for threaded work, but its construction is along the same general lines and brings in some of the same principles, so that it is included in order that the salient features of its design may be noted. The conditions responsible for its design are somewhat peculiar, in that a taper hole — $\frac{1}{8}$-inch taper per foot — has been reamed in the work *A*, and as this piece was made from the bar on a screw machine and then cut off, it is obvious that while the end of the work *B* is approximately square with the taper (as it is faced at the same setting) the other end *C*, where the piece is cut off, will very likely be out of truth. The purpose of the arbor was to hold the work while refinishing the surfaces *D*, *B* and *C*, so that they would run in perfect truth with the taper. The cup-shaped collar *E*, having a three-

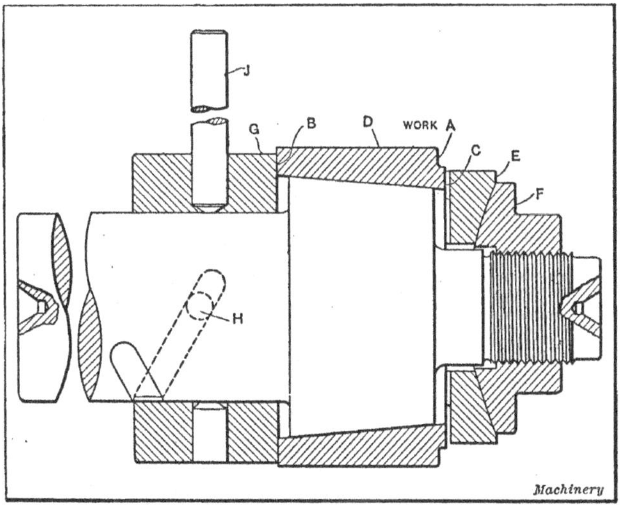

Fig. 19. Male Type of Knock-off Arbor

point bearing against the work, is held in position by the ball-nut F, thereby avoiding any cramping action that would tend to distort the arbor. This nut is released, after the periphery has been turned and the end *B* faced, so that a cut may be taken across the face *C*, thus finishing the piece. As the tapered portion is of such an angle that the wedging action is sufficient to hold the work firmly, no trouble in the matter of driving is experienced, but in removing the work from the arbor, the knock-off collar *G* is brought into use. This collar has an internal pin *H*, working in the right-hand spiral slot shown. The end of the rod *J*, removable at will, acts as a lever by means of which the collar is operated, thus forcing the work from the arbor.

Fig. 20 shows a female arbor of somewhat similar construction to the male arbor which has been noted in Fig. 18. In this case the threaded bushing *A* receives the end of the work *B*. The bushing itself is held in place by the two screws *C*

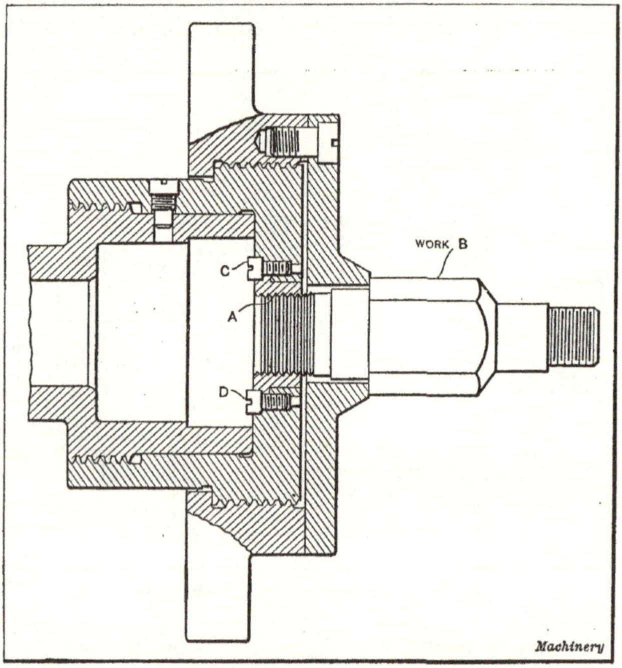

Fig. 20. Female Type of Knock-off Arbor

and *D*, but aside from this difference, the arbor is about the same as the other.

Comparison between Two Designs for the Same Purpose. — It is fully as important, in designing any piece of mechanism, to understand what not to do, as it is to know just what to do. For this reason, an instance is given of an arbor which was not successful, partly because of careless handling when in use on a turret lathe, but chiefly on account of faulty design. The arbor, shown in Fig. 21, was designed for the purpose of turning and facing milling machine arbor nuts, and it was intended to take care of nuts from $\frac{3}{4}$-inch up to and including $1\frac{1}{2}$-inch thread diameter, and of various lengths. The body *A* was made of tool steel, hardened and ground to fit the spindle ring of the turret lathe, and it was drawn up against the shoulder *C* by

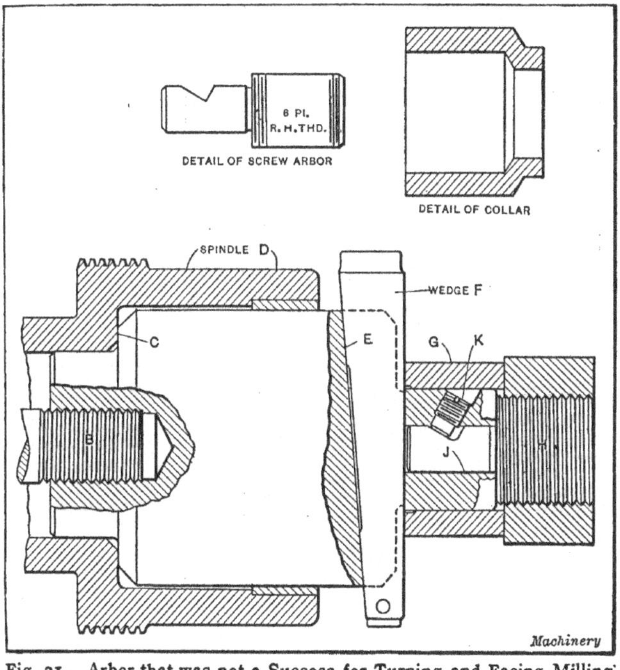

Fig. 21. Arbor that was not a Success for Turning and Facing Milling Machine Arbor Nuts

a rod *B* passing completely through the spindle *D*. The slot *E*, which is tapered on the rear end, receives the knock-out wedge F, and this, in its turn, bears against the slip-collar *G*. The screw-arbor *H*, shown also in detail in the upper left-hand portion of the illustration, was a ground fit in the cylindrical hole *J*, and held in place by the screw *K*. The slip-collars and threaded arbors, shown in detail, were made in various sizes and pitches to suit the nuts which were to be handled.

This entire equipment was unsuccessful, and its use was discontinued after attempting in every way to insure its truth. Although all the parts were very carefully made and fitted, there were certain inherent faults in the design which could not be remedied. After the work had been placed in position and the wedge tapped lightly to make sure that it was firmly seated, it was noted that there was frequently a decided wobble

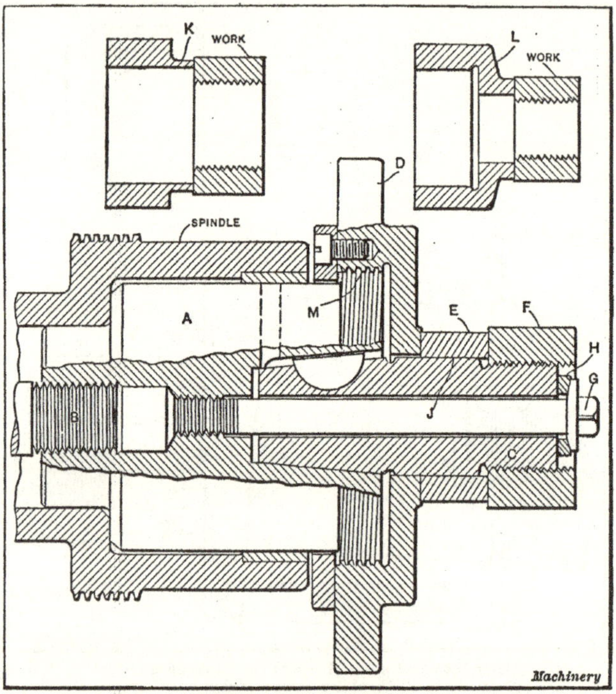

Fig. 22. Arbor designed to replace Type shown in Fig. 21

at the end of the arbor, which should obviously have run in perfect truth. Undoubtedly the action of the wedge had a tendency to throw unequal strains upon the structure, while the screw *K* also helped the matter along, so that the whole arbor was faulty in these particular points, and it was therefore considered a failure.

A decidedly improved construction for the same purpose is

shown in Fig. 22, although its cost was considerably greater than the other. The body *A* in this arbor was located and secured in the spindle in the same manner as the other, *i. e.*, by a threaded draw-back rod *B*, running through to the rear end of the spindle. Every precaution was taken to insure a

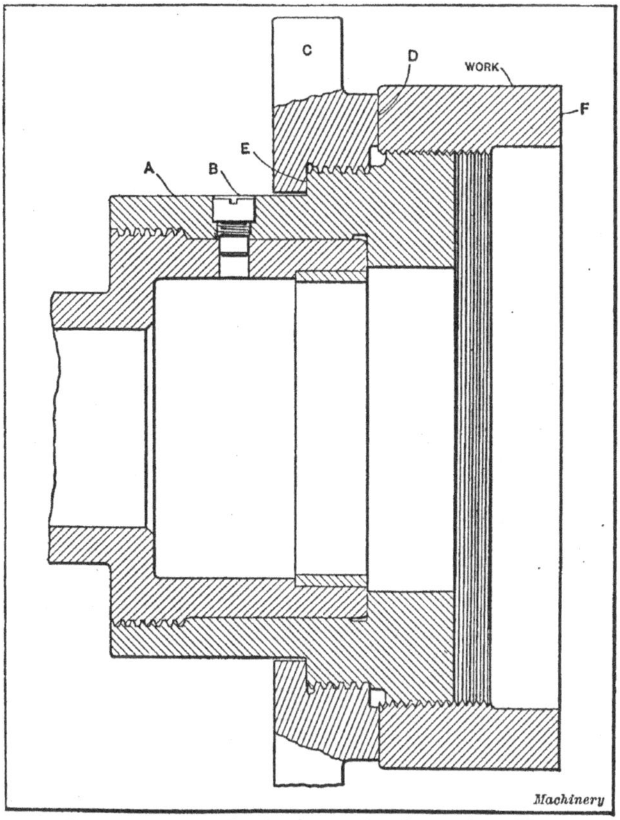

Fig. 23. Knock-off Arbor adapted to Large-sized Work

true running arbor, all essential points and surfaces being ground or lapped in position, in order to avoid all chances for error. The threaded interchangeable arbor *C* was accurately fitted to the conical hole in the body, and was keyed to resist the torsion of the cut.

As in a previous instance the knock-off portion consisted of two parts, shown at *D,* in order to permit assembling while the thread at *M* was 6-pitch left-hand, thus making removal easy. *E* is an interchangeable sliding collar, and it interposes

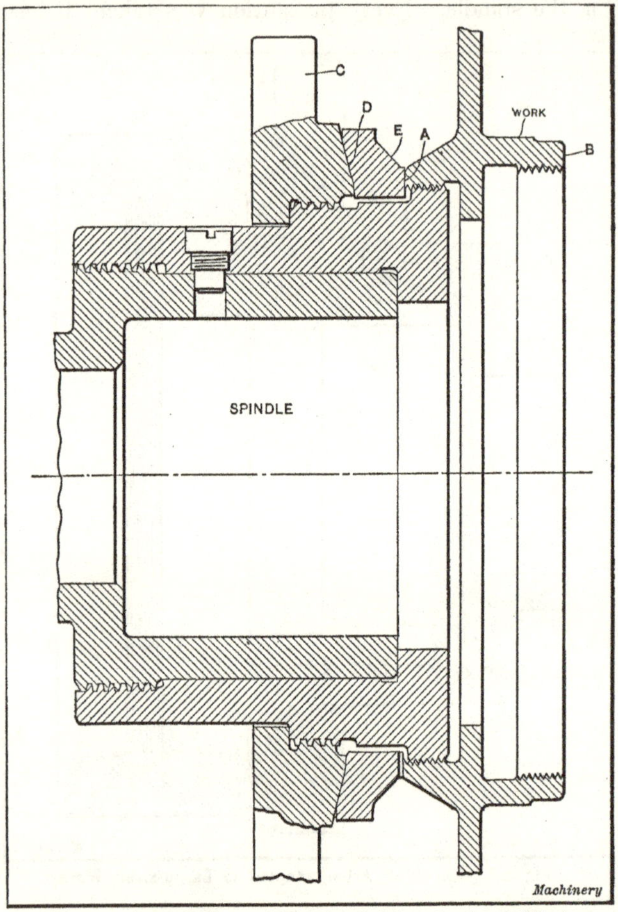

Fig. 24. Another Type of Knock-off Arbor for Large Work

between the shoulder of the knock-off and the work F, obviously acting as a spacer and longitudinal stop at the same time. The ball and cup arrangement of the bolt and washer *G* and *H* serve the purpose of obviating any tendency to "cock" the arbor when it is placed in position. To obtain as rigid a construction as possible, the taper is of generous dimensions and the cylindrical portions of the various arbors are made uniform in size. By referring to the upper left-hand illustration in Fig. 22, it will be noted that the collar *K* is shouldered for a short distance back of the work, to allow the turning tool to perform its function. For some of the smaller sizes a collar like *L* — upper right-hand — was found necessary, due to the difference in the shoulder diameters between *J* and *C*.

Arbors for Larger Work. — When the work is as large as that shown in Fig. 23, a somewhat different construction is necessary. In the case shown, the piece is a threaded steel collar $7\frac{1}{2}$ inches in diameter, in which the end F is to be faced square with the threaded portion. This arbor is applied to the spindle of a large horizontal turret lathe, and its use is obviously restricted to work of a diameter somewhat larger than the turret lathe spindle. The body *A* is screwed directly onto the spindle as a faceplate would be, and it is prevented from turning by the teat-screw *B*. The knock-off portion *C* is threaded with a 4-pitch left-hand thread of the Acme type, running on the body portion, and has the usual projecting lugs for releasing purposes. As it brings up against the shoulder *E*, it forms a positive stop for the work, which locates on the ring at *D*. The action of this arbor is very evident, while its extreme simplicity, it being composed of only two parts, makes it one of the best.

Another large piece of threaded work is shown in Fig. 24, which is considerably harder to handle, in that the ends *A* and *B* are unfinished, and the threads shown must be square and concentric with each other. The construction of the arbor for this piece is similar to that shown in Fig. 23, except that the knock-off portion *C* has the forward face formed to a radius *D*, generated on the center-line of the arbor. The cup-shaped washer *E* has a three-point bearing against the work at *A*, thereby equalizing the strains and preventing distortion so that the end of the work *B* may be bored and threaded in the correct relation to the other end.

Vertical Knock-off Fixtures for Heavy Work. — The application of the arbors previously mentioned has been entirely in a horizontal plane and in the majority of cases, for comparatively small work, but we will now take up some designs of a heavier nature, for use in the vertical turret lathe or vertical boring mill.

Fig. 25 illustrates a heavy knock-off fixture which was designed for use on a vertical turret lathe, and the work shown

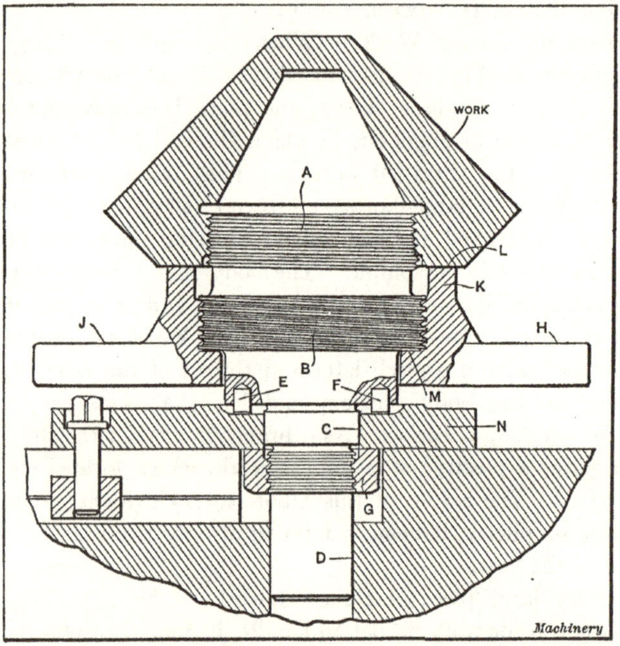

Fig. 25. Knock-off Arbor for Use on Vertical Turret Lathe

in position on the fixture is a large bevel cone of chrome nickel steel, the conical surfaces of which are required to be in perfect truth with the interior threaded portion. This fixture is called upon to resist the strains incident to very heavy cutting and is necessarily of very rigid construction throughout. The steel arbor is threaded on its upper end at *A*, with an 8-pitch right-hand thread, upon which the work is screwed. Below this, marked *B* in the illustration, is another thread of much coarser pitch, *viz.*, 4-pitch left-hand, and with the Acme type of thread. The cylindrical portion *C* passes down through the body of the fixture, and is ground to fit the hole in the table at *D*, thus acting as a locating plug. The two pins *E* and F are obviously used to prevent the arbor from turning, while the nut *G* draws it down against the shoulder. It may be noted that the fixture body *N* is firmly secured to the table by the bolts shown, these being screwed into shoes in the table T-slots. The knock-off portion is a steel forging, having an upper cylindrical section *K* which bears against the work at *L*. A positive vertical stop is assured by the shoulder *M* on the arbor. When using the fixture, the knock-off forging is always brought up against this shoulder before the work is screwed onto the arbor. After the piece has been machined, a sharp blow is given to either of the two lugs *J* or *H*, thus reducing the friction at *L* and permitting the easy removal of the completed cone.

Fixture for a Large Taper Thread. — Another vertical fixture which requires in its design the solution of the most difficult problems in the construction of knock-off arbors and fixtures is illustrated in Fig. 26. The work is a large steel forging, a detail of which is shown in the lower part of Fig. 16, and it will be noted that the threaded ends are pipe thread taper and must be in line with each other. To make the conditions worse, the end of the hub is unfinished, necessitating careful handling and clamping in order to avoid distortion. In order to understand the situation thoroughly, it may be stated that, in a previous setting of the work, a straight hole has been bored entirely through the piece, and the taper hole on the hub end has been reamed and threaded to the correct pipe thread taper. The taper bushing *B* is threaded on the outside to the pitch of the pipe thread, and it is screwed in place in the work before placing it in position on the fixture. Two slots in the end, shown at *C*, permit the use of a spanner. The interior cylindrical portion is ground to an easy running fit at *D*, and is threaded somewhat loosely on *E*, which is a 6-pitch Acme right-hand thread. The centering and alignment are governed entirely by the upper cylindrical bearing.

The arbor itself has a 4-pitch left-hand Acme thread cut upon it at *F*, and it is driven through the square keys *G* and *H* in the body of the fixture. As the lower end of the work is unfinished, it is necessary to support it in such a way that it will

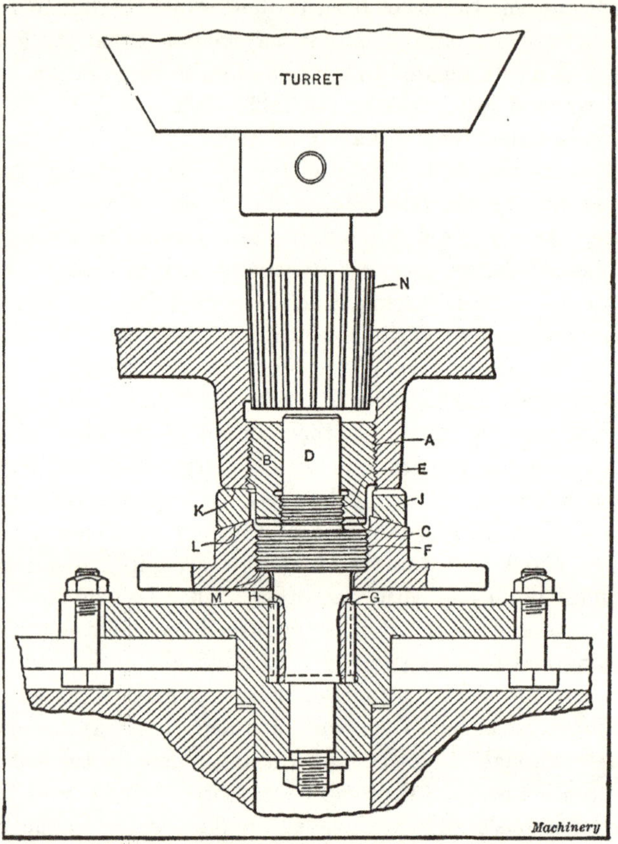

Fig. 26. Another Style of Vertical Knock-off Arbor

not be distorted, or thrown out of the vertical position. To obtain this result, the collar *J* must be designed to give a three-point bearing to the piece on the surface *K*. The collar is therefore slightly relieved leaving three high points which bear against the hub, and the cup-shaped bottom allows a rocking action upon the spherical portion of the knock-off at *L*. Except for the spherical surface, this knock-off is about the same as that shown in Fig. 25. Attention is called to the manner in which the positive vertical location of the work is assured, irrespective of the tapered portion, as this is one of the important features of the fixture. It will be readily seen that the clearances between the bushing *B* and the shoulders on the arbor are sufficient to allow for considerable variation in the depth of the tapered hole, and still keep the relation between the surfaces *M* and *K* constant. The taper reamer shown at *N* helps to make clearly apparent the actual working conditions while the fixture is in use.

The majority of the common conditions have been noted in the illustrations shown in this chapter, and while more difficult problems may arise occasionally, they may be successfully handled by adaptations of the foregoing.

CHAPTER VI

HOLDING DEVICES FOR LATHE AND BORING MILL WORK

Chucking Fixtures for First-operation Work. — The methods of holding and clamping rough castings for the first or "chucking" operation are so diversified that the subject must, necessarily, be treated by means of examples representing different varieties of work. Nearly all of the examples shown are more or less cylindrical in shape, for the reason that elliptical, rectangular, or odd-shaped parts require special treatment and, therefore, will be treated separately in a subsequent chapter. In the general course of manufacturing, there are occasionally pieces of peculiar shape which require chucking fixtures, but as this work is of such great variety, it is difficult to give much information regarding its handling except in a general way. Any piece of work of peculiar shape requires a thorough knowledge of the conditions governing its use, in order that it may be chucked properly and located from the surfaces which are of the greatest importance.

Important Points in the Design of Chucking Devices. — In the design and construction of chucking devices, there are a number of points to which the most careful consideration must be given. In some cases, the work must be held by the cored interior, as, for example, an automobile piston, or, in fact, any other work in which it is necessary to have an equal division of metal throughout the cylindrical walls. In other instances, however, some method of exterior holding may be perfectly satisfactory. The term "exterior holding" does not necessarily mean that chuck-jaws are referred to, for various devices, other than jaws, will be cited during the following discussion of holding methods.

Having determined whether the work is to be held externally or internally, let us take up the important points in the design of holding devices.

1. The important locating surfaces should be carefully considered, always having in mind the future handling of the piece in its various operations. Great care should be taken that no locating points are so placed that they will come in contact with the work in places where the pattern is gated, or where numbers or letters may appear.

2. In setting up a rough casting there should never be more than three fixed supporting points; any others which may be necessary for the proper support of the work must be made adjustable, with some approved method of clamping securely after adjustment.

3. The work must be firmly secured so that no distortion can take place under the strain of clamping.

4. When the work is of such a nature that difficulty is experienced in obtaining proper clamping surfaces, it is sometimes advisable to consult with the patternmaker in regard to the addition of clamping lugs to the pattern. In cases of this sort, these lugs should be so applied that they can be easily removed later.

5. In designing a chucking fixture the safety of the operator should be considered carefully; protruding heads of screws, bolts, clamps and similar parts should be avoided as much as possible.

6. Convenience and accessibility in setting, locating and clamping the work are also of primary importance.

Individual points regarding the work-holding devices shown in the illustrations will be discussed. We shall consider holding devices for the horizontal turret or chucking lathe, the vertical turret lathe, and the vertical boring mill. In describing these devices, the work and its requirements will be considered, as well as the important locating surfaces, the method of handling the work and important points in the design of different fixtures.

Two-jaw Chuck arranged for Internal Chucking. — It is essential that the cast-iron piston shown in the upper part of

Fig. 1 be located from the cored interior, in order to have the outer walls concentric with the core, thus obtaining an equal distribution of metal throughout the piston walls. Due to the formation of this casting, the core is poorly supported at the closed end, and, therefore, has a tendency to drop slightly when the metal is poured into the mold, thereby producing a lack of concentricity between the cored portion and the exterior surfaces. It is logical, then, in order to obtain uniform results, to work from the cored portion when setting up the casting. Many methods of holding have been devised for this purpose.

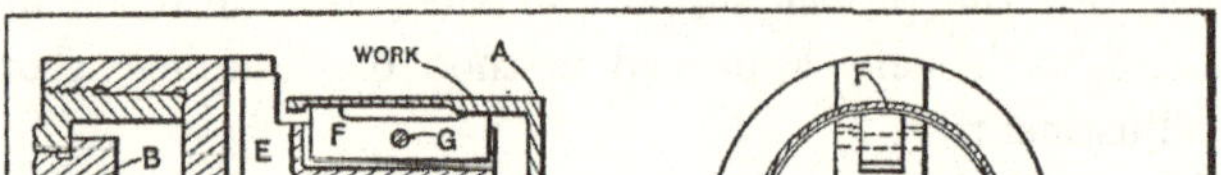

Fig. 1. (Upper View) Two-jawed Chuck for Holding Piston internally; (Lower View) Chucking Fixture for Piston Ring Pot

The fixture shown in Fig. 1 is one of the simplest, an ordinary two-jaw chuck with special internal jaws being used for holding and locating the work *A*. A steel bushing *B*, fastened into the spindle, contains the stop-rod *C*, which comes against the head of the piston, thus insuring a uniform thickness of metal at this point. The chuck is supplied with the two special jaws *D* and *E*. The former is a plain jaw with two bearing points, while the other has a swivel-jaw *F*, pivoted on the pin *G*, which allows it to conform to the inequalities of the casting. This method of chucking is one of the cheapest, and the results obtained by its use are fairly satisfactory. There is a tendency toward inequality in the thickness of the piston walls in the direction of the wrist-pin bosses, due to the fact that the centering action of a chuck of this type is in two directions only; however, at least one large manufacturer in the East uses this method entirely. The chuck is employed for rough-turning only, thus securing a partially finished surface which is true with the core and which may be used to work from for subsequent operations.

The work *H*, shown in the lower view of Fig. 1, is a cast-iron piston ring pot, which must be held in such a way that it can be bored, turned eccentrically, and separated into narrow rings for a gas engine piston. As the ring pot is very thin, it must be carefully held to avoid distortion and yet be very rigidly secured, as there are several tools working at one time so that the torsion produced by the cut is excessive. The pot is made with an internal gripping ring *K*, which is slightly beveled to assist in keeping it back against the chuck jaws. The chuck is an ordinary three-jaw, geared-scroll type, having jaws as shown in section at *N*. These jaws are of steel and are drilled to receive the hook-bolts *L* which pass entirely through them and grip the ring from the inside. The heads of the bolts *M* come out through slots in the jaws, the heel having a backing at *O*. When setting a casting the bolts are left free while the jaws are brought up against the outside of the casting with just enough pressure to get a bearing. The bolts are then set up tightly on the gripping ring, so that the work is held firmly but without distortion. This method is very good and can be applied successfully to many varieties of thin work. The hook-bolts are of tool steel and are hardened and drawn to a deep straw on the hook end. The backing up of the hook-bolt at *O* is very important, for unless properly supported at this point its action is greatly impaired and it soon becomes bent out of shape and is absolutely useless.

Ring Pot Locating Fixture without Chuck-jaws. — Another cast-iron ring pot of somewhat different form is shown in Fig. 2.

The operations on this piece are identical with those for the casting shown in the previous illustration, but the holding method is entirely different. This form of pot is used by one of the largest manufacturers in this country. Before it is placed in the fixture, the face of the gripping ring *A* is ground square with the outside of the pot. The body of the fixture *B* is of cast iron and is screwed fast to the spindle nose of the horizontal turret lathe upon which it is used. The annular ring

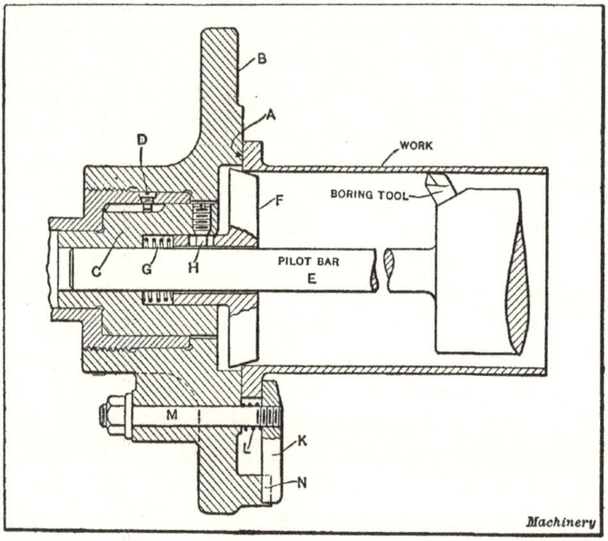

Fig. 2. Another Ring Pot Chucking Fixture

or pad against which the ring pot lies is faced square in position on the machine. A hardened and ground tool-steel bushing *C* is accurately fitted to the inside of the spindle and is held in position by the teat-screw *D*. It will be noted that this bushing also acts as a guide for the boring-bar pilot *E*. A tapered plunger *F* is forced outward by the spring *G* and centralizes the inside of the pot. The screw *H* simply acts as a retainer to keep the plunger in position. There are three clamps, 120 degrees apart, on the face of the fixture. One of these is shown at *K*; obviously this is tightened by the screw *M*, while the coil spring *L* serves to keep the clamp away from the work when not in use. The lug *N* prevents the clamp from twisting around when the screw is being tightened. This fixture is a very good one, except that its operation is rather slow.

Locating Fixture for a Ball-and-socket Pipe Joint. — The requirements for the work shown in Fig. 3 need little explanation. The piece itself is a steel casting. A cast-iron "cat-head" or fixture *A* is screwed onto the spindle nose, and is faced at *B* to an arc corresponding with the rough ball-portion

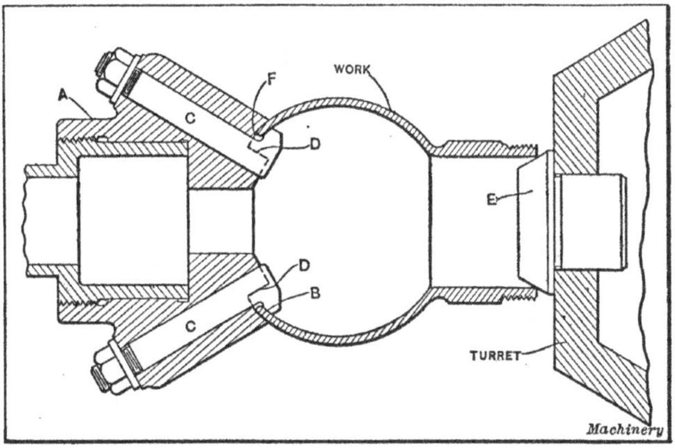

Fig. 3. Fixture for Ball-and-socket Pipe Joint

of the pipe joint. The two hook-bolts *C* obviously grip the work from the inside and hold it firmly against the finished face *B*. A centering plug *E* fits the turret hole and is brought up and entered into the casting before the hook-bolts are set up tightly. This fixture was not entirely satisfactory owing to the condition of the rough castings at the end *F*, for at this point they varied greatly and were very rough, making the holding somewhat uncertain. A method of holding this work by the interior undoubtedly would have been more satisfactory.

Chucking Device having an Outboard Supplementary Bearing. — The automobile tail-shaft housing shown at *B* in Fig. 4 is made of malleable iron and is so long that chucking by means of jaws is out of the question, on account of the excessive overhang which would be necessary. The piece was to be finished complete in one setting and the fixture shown was designed and used for this purpose. The body *C* is of cast iron and is screwed onto the spindle nose. The inner cylindrical surface *D* is very carefully bored and the outer bearing surface *E* is turned and finally lapped to a nice running fit in the bracket *A*. The periphery *H* of the locating and centering bushing *F* is crowned on a radius and is slotted in various places, as shown at *K*, to receive the exterior ribs on the housing. A pointed set-screw *G* keeps the bushing in position. The tapered plug *L* is located in the turret hole and serves to center the work, and the pointed set-screws *M* (three of which are used) are sunk into the casting and act as drivers in addition to holding it in the position determined by the tapered plug. The bracket *A* (also shown in detail) acts as an outboard bearing for the long body of the fixture and prevents the vibration which would otherwise result from the excessive overhang. A glass oil cup was an added refinement to the equipment and may be noted at *N*. This fixture has been used with great success by a large manufacturing firm.

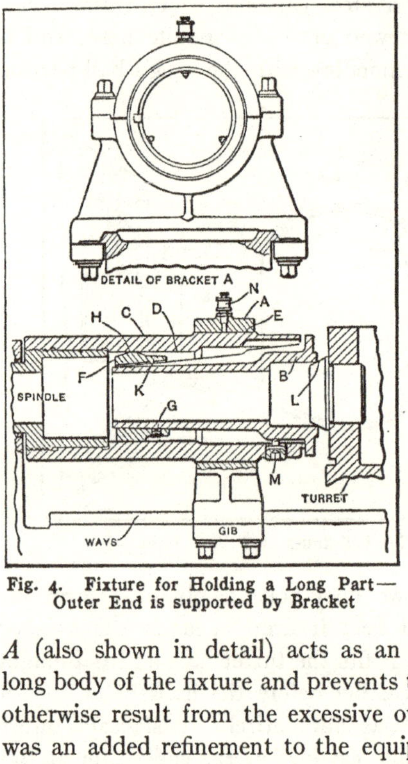

Fig. 4. Fixture for Holding a Long Part—Outer End is supported by Bracket

Method of Chucking One-half of a Rear Axle Housing.—The male portion of an automobile rear axle housing is shown in Fig. 5. This is machined in one setting on a horizontal turret lathe. The body of the fixture *B* is of cast iron and is screwed onto the nose of the spindle. Three steel pins *C* are located 120 degrees apart, around the inside of the fixture body, the coil springs *D* forcing them outward and the set-screws *E* securing them in place when properly located against the work. The method of clamping is somewhat peculiar and should be carefully noted. The swinging dogs F have a knife-edge at *K* and are pivoted on the pins *G*, which are set back in such a position that the action of the dogs (controlled by the hollow set-screws *H*) has a tendency to carry the work back against the body of the fixture and the spring jack pins. A steel bushing *L* is forced onto the small end of the work and assists in centering it in the spindle. This bushing is crowned on a radius the same as that shown in Fig. 4. The taper locating plunger *N* is forced out by the spring *O* and is restricted in its action by the pin *P*. The two-arm support *Q* is of cast iron and is of assistance in keeping the work in position while the various screws and dogs are being tightened. The bushing *M* acts as a guide for the boring-bars and reamers used in machining the work. The work is driven by the ribs *R*, which enter slots in the body of the fixture. This method of holding gave satisfactory results, although considerable care was necessary to avoid springing the casting, when tightening the clamping dogs.

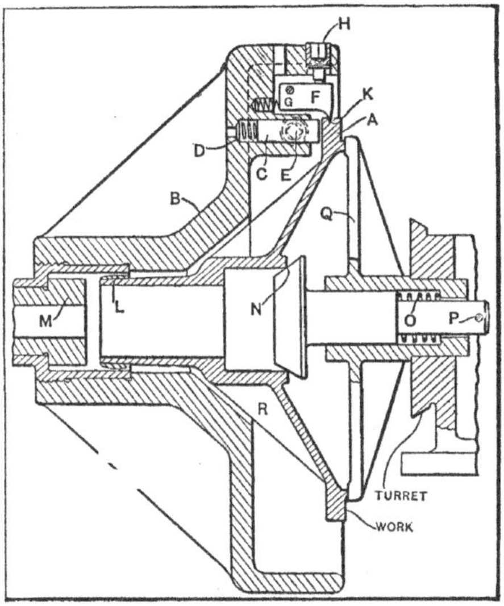

Fig. 5. Casting A held so that it can be Machined at one Setting

Equalizing Pin Chuck for a Gas Engine Piston. — One of the many varieties of internal holding piston chucks is illustrated in Fig. 6. Although rather expensive, it is an excellent example of this type of chuck, and is very well made. All working parts are of steel or bronze and all parts requiring such treatment are carefully hardened and ground. The body of the chuck is of machine steel, carbonized, pack-hardened and ground; the pins, cams, operating rod, screws and bushings are of tool steel, while the miter gears are of bronze. The body of the chuck *A* is screwed onto the spindle nose and is ground or lapped at all important points. The operating cams *B* and *C* are slotted in three places around the periphery at *D* and *E*, these slots being angular and forcing the six pins F and *G*

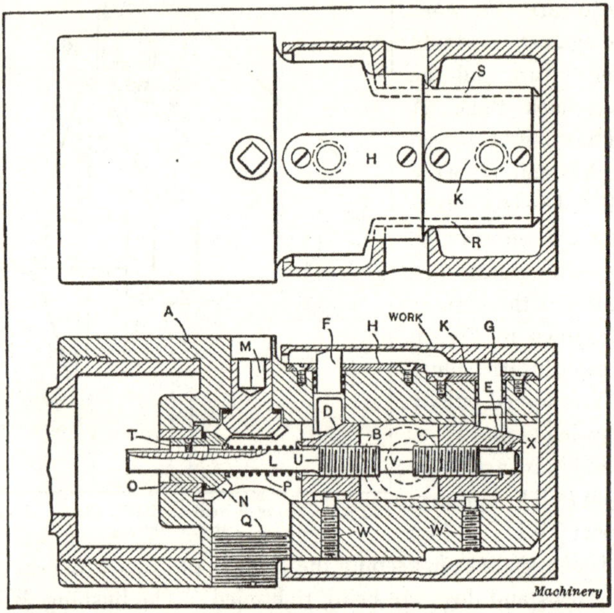

Fig. 6. Special Chuck for Holding Gas Engine Pistons internally

out against the interior walls of the piston. It may be noted that the steel plates *H* and *K* are let into the body of the chuck, and act as retainers for the pins. These plates are clearly shown in the upper view. The operating rod *L* is revolved through the action of the miter gears *M* and *N*. The latter has a key *T* engaging a long spline in the operating rod, which is thereby permitted to move longitudinally. The threaded portion *U* has 6 threads per inch, right-hand, while that at *V* has 6 threads per inch, left-hand. The forward cam is packed with felt at *X* to keep out the dirt. The bushing *O* is of tool steel, hardened and ground. The plug *Q* simply closes the hole which has been put in for assembling purposes.

By referring to the upper view it will be seen that the chuck body is cut away on the sides at *R* and *S*, on account of the wrist-pin bosses in the piston, and the overhanging lip at *R* acts as a driver. In designing a chuck of this kind, it must be remembered that while the rear clamping pins may be equally spaced, the position of the forward pins will be determined by the diameter and spacing of the wrist-pin bosses, and an end view will be found essential to determine the correct position. In general it will be found that two of the forward pins seldom can be spaced more than 80 degrees apart and often the spacing cannot be made more than 55 or 60 degrees. Another point in design which is of great importance is the amount of clearance between the ends of the wrist-pin bosses and the flattened sides of the chuck body. It is seldom safe to allow a clearance of less than $\frac{3}{16}$ inch on each side, over the finished sizes called for on the drawing of the piston. The location of the stop-pins *W* is also important, and sufficient allowance should be made in the length of the cam slots to take care of variations in the piston castings. A chuck of this type gives results which are satisfactory in every respect.

An Equalizing Pin Chuck for an Electric Generator Frame. — The examples which have been referred to in the foregoing are all adapted for use on the horizontal type of turret lathe, but we shall now go a step farther and take up chucking devices designed for the vertical turret lathe and the vertical boring mill. As machines of this type are adapted more to heavier classes of work, the fixtures should be designed with relation to the work and also to the power of the machines upon which they are to be used. For machining the steel generator frame shown on the fixture in Fig. 7 the working points specified by the manufacturer are at *B* between the ribs *C* on the upper portion of the casting *A*. It was further specified that the work must be held by the core to insure an evenly balanced casting.

The design of this chuck resembles that shown in Fig. 6, in that both chucks are fitted with pins and operating cams; the operating mechanism, however, is entirely different. The body of the chuck *D* is of cast iron; it is carefully reamed and lapped at important points, and is securely fastened down to the table of the machine by three screws having T-shaped heads, which enter the table T-slots as shown at *E*. The fixture is centered by the hollow stud *F*, which is of tool steel, hardened and ground inside and out. This stud also acts as a bushing for the operating sleeve *G*, and is cut away for clearance on one side where the spiral gear *H* passes through it. This gear meshes with another which is cut on the outside of the operating sleeve. The latter is of bronze and it is threaded internally with a 6-pitch Acme thread, corresponding with that cut upon the lower end of the cam *K*. The operating rod *L* is pinned into the upper cam *M*, and is shouldered and journaled at its lower end where it passes through the operating sleeve at *N*. The coil spring *O* is so proportioned that it simply supports the cam mechanism and assists in releasing. It will be noted that the arrangement of this internal mechanism permits a "floating" action for the cams, so that the clamping pins all bear uniformly against the inner walls of the casting.

In setting up the work on the fixture, it is dropped over the top until the lower end rests on the three adjusting jacks *P*, which are placed 120 degrees apart and are knurled for finger adjustment. The upper locating arms *B* are swung back out of the way while the casting is being set in position, but as soon as this has been done they are brought around into the position required. The jacks are next raised until the casting has been properly located against the arms; then a long-handled socket wrench is inserted at *Q* and the gearing is revolved until the pins are securely seated against the inner walls of the work.

A Combination Device having Equalizing Pins and Hook-bolts. — The piece shown at *A* in Fig. 8 is a clutch pulley for a gasoline tractor. It is made of cast iron and the method of

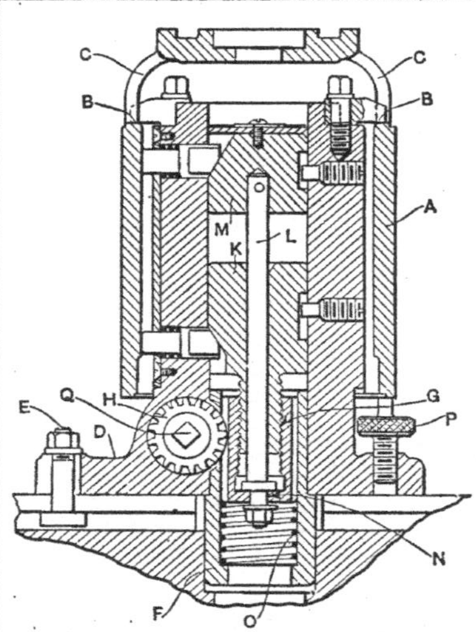

Fig. 7. Vertical Boring Mill Fixture for Holding Part A internally

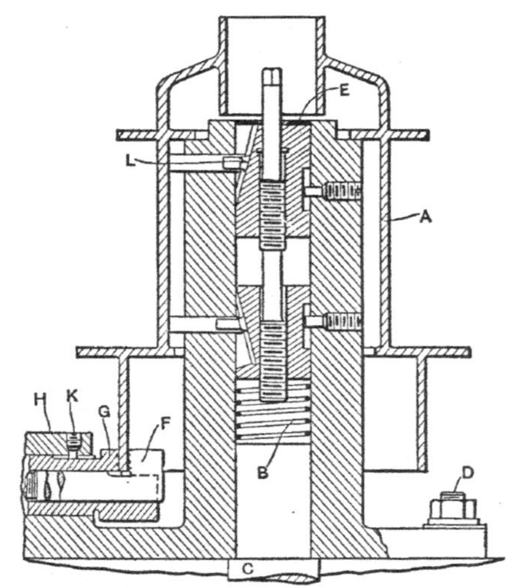

Fig. 8. Vertical Fixture having Internal Clamping Pins and Hook-bolt Clamps for Lower Flange

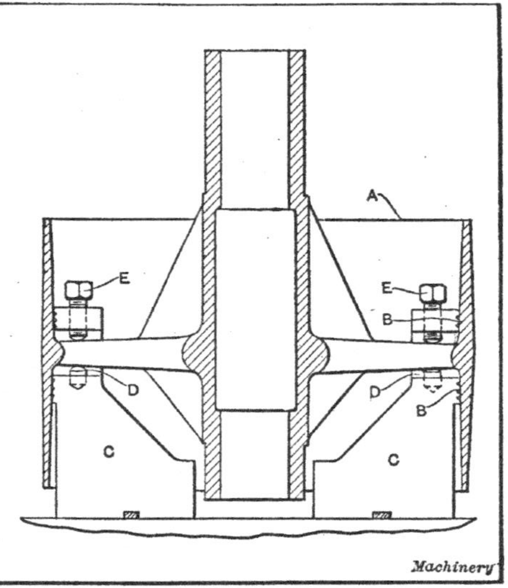

Fig. 9. Special Chucking Jaws for Large Pulley

holding from the inside was decided upon because it seemed to offer better facilities for machining. As in a former example, the body of the fixture contains the cams and the operating rod which is threaded right- and left-hand, as before. The lower ends of the pins and the slots in the cams are dovetailed in this instance, so that the outward and inward movements are controlled mechanically, no springs or plates being required. The fixture is centrally located on the machine table by the plug *C*, which fits the center hole in the table and is held down in the usual manner by the T-bolts shown at *D*. The coil spring *B* simply acts as a support for the cams and rod. An annular groove is cut in the upper cam at *E* and this is packed with felt to assist in keeping the dirt out of the working parts.

The lower part of the fixture has three bosses (one of which is partially shown at *H*), which contain the floating jaws and hook-bolts, *G* and *F*, for clamping the lower flange of the pulley. The construction of these parts is more clearly shown in Fig. 10, and, as they are identical, the reader is referred to the portions marked *C* and *D* in that illustration. The results obtained by the use of this fixture were at first very satisfactory, but, after a time, the dirt which gradually accumulated in the dovetail cam slots, began to cause trouble, until, finally, it became almost impossible to operate the mechanism. Then, too, in several cases the dovetail part broke off completely, necessitating new pins, so that the chuck as a whole cannot be considered an absolute success. No trouble would have been experienced if the cam slots and pins had been made as shown in Fig. 7, with coil springs and retaining plates.

A Set of Special Jaws for a Large Crowned Pulley. — The large farm engine pulley shown at *A* in Fig. 9 is of cast iron, and it must be held by the inside in such a way that it will not be distorted while fairly heavy cutting is being done on the periphery of the pulley. A four-jaw table was selected on which to hold the work, as there were eight spokes in the pulley. The chuck jaws *C* were made of 0.40 per cent carbon steel and were slotted for the spokes as shown in the illustration. The hardened steel studs *D* were set into the slots, and the set-screws E brought down tightly on them after the jaw surfaces B had been brought out to center the pulley. This method of holding pulleys or other spoked work gives very satisfactory results and is used by many manufacturers throughout the country.

A Chucking Device for a Difficult Piece of Electrical Work. — The aluminum piece shown in Fig. 10 is one of the most difficult

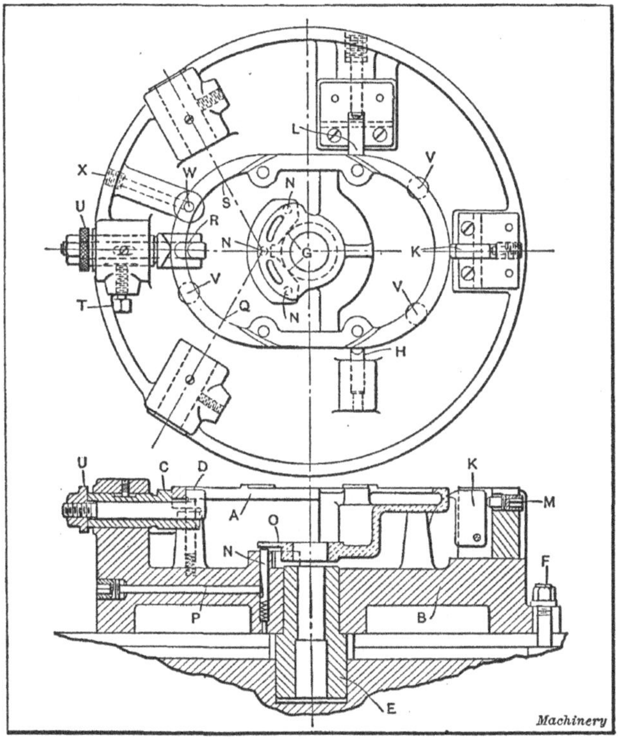

Fig. 10. Ingenious Chucking Device for Holding Odd-shaped Aluminum Casting without Distortion

for which the author has ever had occasion to make a chucking fixture. The walls of the entire piece were of very thin section, and the overhanging portion A was elliptical in shape and entirely unsupported. It was necessary to so hold this casting that it could be machined without distortion. The body of the fixture *B* is of cast iron. It is centrally located on the table by means of the hollow bushing *E*, and clamped down by T-bolts shown at *F* in the table T-slots. A portion of the fixture shown in the upper view at *G* was cut out to form a V, in which the cylindrical portion of the casting is centered. One of the sides of the casting is located against the knife-edged pin *H* (shown in the upper view), and the casting is forced into the V and against this knife-edged surface by the swinging clamps *K* and *L*.

It will be noted that these clamps also have a knife-edge and that the pins upon which they are hung are in such a position that their action is downward, thereby tending to hold the work securely against its supports. Hollow set-screws *M* control the action of these swinging clamps. There are three spring-pins *N* which are used to support the very thin flange *O*, and the springs which force them upward are carefully proportioned so that they have just sufficient strength to insure contact without springing the work. These pins are locked in their positions by the long hollow set-screws *P*, shown in the lower view.

One of the principal points of interest in this fixture is the method of gripping the overhanging elliptical portion *A*, without distorting the casting. This is accomplished by means of the floating jaw *C* and the hook-bolt *D*. There are three of these floating clamps which grip the work at points *Q*, *R* and *S* (see plan view). It will be noted that the clamps are free to "float" laterally, until the set-screws at *T* are tightened. The knurled hexagon nut *U* is threaded onto the head of the hook-bolt, and is lightly tightened by means of the fingers before the final tightening with a wrench. As these clamps have a perfectly free floating action, until the binding screws are tightened, obviously there can be no distortion of the piece, and yet it is held rigidly at these points. When the work is placed into the fixture it rests upon three fixed points shown in the plan view at *V* and a fourth point *W*. The latter is held upward by a light spring and it is locked in position by the long hollow set-screw *X*. The work accomplished by the use of this fixture was true and accurate, no evidence of springing out of truth being apparent.

Boring Mill Fixtures for a Large Ball Joint. — The work shown at *A* in Fig. 11 is a large ball pipe joint for a suction dredge, and is to be faced on the upper surface *B*. Two designs were made for this work. The special jaws *C* and *D* are bored to an arc corresponding with the ball, and the work rests upon and is centered by these jaws. The body of the clamping device *E* is of cast iron and is bolted down to the table. It contains four pointed screws F, which, normally, are kept away from the work by means of coil springs. The ends of the screws are beveled to the same taper as the operating screws *G*, which are threaded at their upper ends with a coarse pitch thread, and have squares to receive the end of the long socket-wrench *H*, by means of which the screw is revolved, thus causing the points to sink down into the casting. Wood braces are used to support the flange which is to be faced. Four are used, although only two (*K* and *L*) are shown in the illustration.

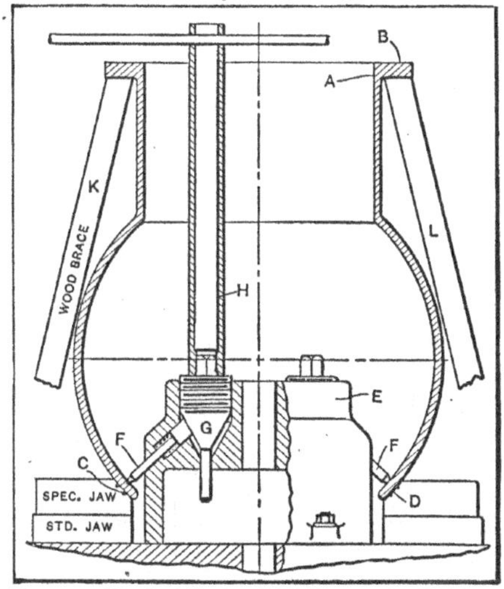

Fig. 11. Vertical Boring Mill Fixture for Holding Large Ball Joint

Fig. 12 illustrates another fixture for holding the same piece of work, which gives more satisfactory results than the one shown in the previous illustration. It can be manipulated more rapidly, because it is more accessible to the operator. Four sets of special jaws *A* are used for holding the work. The part *B* is formed to an arc corresponding to that part of the work which it grips, and has teeth to assist in obtaining a firm "bite." The supplementary sliding jaw *C* is forced into position by the set-screw *D*, thus clamping the work tightly between *B* and *E*. Wooden supports are used under the flange.

Chucks and Fixtures for Second-operation Work. — In the majority of cases, work which is to be handled in more than one setting is so treated in the first of these that an interior surface can be used from which to locate the piece in the second setting. For work of this nature some kind of internal holding device, such as an expanding arbor or locating plug with clamps, is conceded to be of the greatest utility. (See Chapter V.) Frequently there are instances, however, which require an entirely different method of holding, namely, by some form of contracting device such as a set of collet jaws, a step chuck, soft jaws or some other scheme of a kindred nature. Obviously, one of these devices would only be found necessary when an outside finished surface was to be used as a locating point.

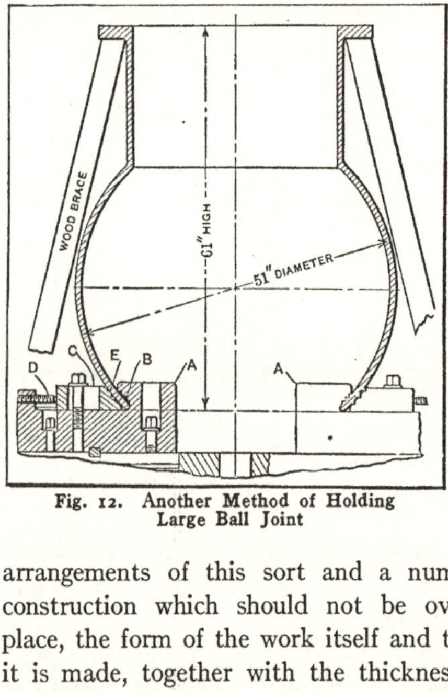

Fig. 12. Another Method of Holding Large Ball Joint

There are a number of conditions which affect the design of arrangements of this sort and a number of vital points in construction which should not be overlooked. In the first place, the form of the work itself and the material from which it is made, together with the thickness of the walls used as locating surfaces, are prominent factors which have a strong influence on the method of holding. The size and weight of the casting naturally determine to a certain extent the type of machine most suited to the work. The degree of accuracy required in the finished piece is also a factor in determining the type of holding device to be used. The types of machines most frequently employed on work requiring fixtures of this kind are the horizontal screw machine or turret lathe, the vertical turret lathe, and the vertical boring mill.

Points in Design. — A few of the most important points in the design and construction of this class of fixtures are given in the following paragraphs:

1. Decide on the method of holding which will not produce distortion in the finished work. Consider the thickness of the metal at the holding points, and note whether there is danger of crushing if considerable pressure is applied. Use a driver of some kind whenever possible, as this will be of great assistance in taking the thrust of the cut, and will therefore do away with the necessity for excessive pressure in holding. It may be permissible in some cases to drill a hole in the work for the express purpose of furnishing a place in which a driver can be inserted.

2. Decide on the type of machine to be used, selecting that most suited to the work, taking into consideration the diameter and weight, and the speed at which it should be run to secure maximum production. In this connection it is well to bear in mind that the diameter to be cut determines the speed, irrespective of the outside diameter of the work itself. For example, a comparatively small hole in a hub might be machined when the outside diameter requires no machining whatever at this setting. In a case of this sort it would be advisable to select a machine capable of sufficient speed to produce maximum cutting speeds at the point where the cut is to be taken.

3. Rapidity of operation is important; this requires accessibility of various clamps, screws, or other adjustable portions necessary to the proper holding of the work. If a number of nuts are to be tightened, see that they are made of such a size that one wrench can be used for all of them. It is a very good idea to make nuts of the same size as some of the nuts on the machine itself, so that the regular wrench which goes with the machine can be used.

4. Provide for positive location for the work, so that variations will not occur in shoulder distances when the piece is completed. It is well to make the locating points conveniently accessible so that they can be easily kept free from dirt and chips. Separate points, pads or studs should be used for locating, whenever possible, rather than unbroken surfaces.

5. The accuracy required in the finished work should be carefully considered. When very close work is called for it is well to make suitable provision for wear in the moving parts and also to allow for grinding or lapping when making the fixture.

6. Rigidity of design is important. It is well to remember that the closer the work can be held to the end of the spindle the less the overhang will be. Less overhang means less liability to chatter and also greater rigidity. Extra strength should be provided at points where there is excessive strain, and ribbing may be put in where needed. The balancing of the fixture should also be looked into, especially if it is to run at high speed.

7. Safety of the operator should be considered and care taken in the design to provide guards over projecting lugs, set-screws, etc. Various methods will suggest themselves to the designer during the progress of the drawing, and a little thought on this point will be greatly appreciated by the operator. Projecting set-screws, exposed gears, etc., are growing more and more out of favor in machinery, and it is time that such things were done away with on fixtures also.

8. The cost of the fixture should bear a certain relation to the number of pieces for which it is to be used; by this is not meant that it should be in a certain exact proportion to it, but that it should be taken into consideration while designing, so that the cost will not be too great as compared with the number of pieces to be machined. For example, if a fixture is to be used for only a hundred pieces, it should be designed as cheaply as possible, so that when the fixture cost is distributed among these pieces it will not increase the cost

of production any more than can be helped. On the other hand, let us suppose the fixture is to be used for a lot of 10,000 pieces. It is then evident that an elaborate fixture having all possible provisions for rapid clamping, etc., would be in place, for the cost distributed among this number of pieces would be very small and the time saved would more than make up for the extra cost.

Outside Holding Devices for Small Work. — There are several methods by which small work can be successfully machined when a second setting is necessary in which the work is held by the outside previously finished surface. The device most used is probably some form of collet jaws, such as is shown in the upper portion of Fig. 13. For second-operation work on bar stock, a method of this kind gives very satisfactory results and is too well known to require a detailed explanation. The collet jaws are closed in on the work either by drawing them back into a tapered body or by forcing a tapered body over them, thereby causing them to contract; the operating lever or wheel is at the rear end of the spindle. In the case shown in the illustration, the second of these methods is used. The machines supplied with this type of collect mechanism are the horizontal turret lathes made by the Pratt & Whitney Co., Hartford, Conn. A noteworthy feature of this collet chuck is the fact that the operation of the closing mechanism does not tend to move the stock lengthwise. It is therefore especially valuable for second-operation work.

Another method employing a step-chuck is shown in the second illustration of the same figure. This method is often used for second-operation work on gear blanks or pieces of a similar nature. The steel blanks *B* can be stepped out for several sizes, as indicated. The closing operation is performed by means of the collet mechanism, and it may be noted that, as in the case of the collet jaws, there is no longitudinal movement of the work. This method is less frequently used than the collet jaws, but it is very useful for many kinds of comparatively small pieces. This device is also a product of the Pratt & Whitney Co.

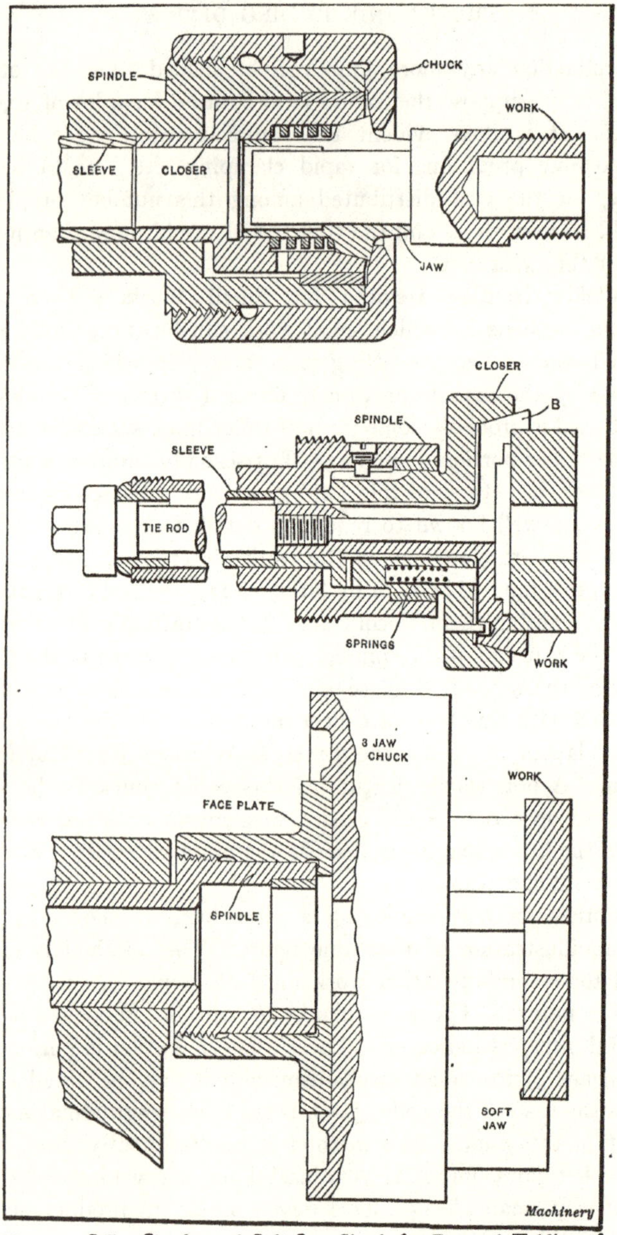

Fig. 13. Collet Chucks and Soft Jaw Chuck for External Holding of Small Work

One other very widely used arrangement is shown in the third illustration, Fig. 13. This is a set of soft steel or cast-iron jaws which are bored out to the proper size, and then clamped onto the work by means of the regular chuck mechanism. Jaws of this kind are used for a great variety of work, both small and large. They are adapted for use in two-jaw, three-jaw and four-jaw chucks, and are made up in a great many forms to suit peculiar conditions. In the two-jaw variety they are frequently formed to fit some odd-shaped piece. In the three-jaw type they are usually simply bored out and shouldered; the four-jaw style is not as common as the other two. In the operation of a four-jaw independent chuck with soft jaws, two of these are left set while the other two are used for clamping. One of these methods is in use in every shop or factory in the country.

Holding Device using a Three-jaw Chuck with a Floating Scroll. — The work shown at *A* in Fig. 14 is an alloy steel jack-shaft gear for an automobile, part of which has been machined both inside and outside in a previous setting. As the size of the hole was too small to permit the use of an expanding arbor, the arrangement shown was decided upon. A regular three-jaw geared scroll chuck *N* of a standard make was selected for the work. The body of the chuck was bored out at *H* and *J* to a dimension $\frac{1}{16}$ inch larger than the width of the scroll ring *G*, thereby permitting the scroll to float radially. The revolving motion of the scroll is accomplished in the usual manner by means of the pinion *L* which is squared at the end *M* to receive a socket wrench. The bushing *K* acts as a retainer. The radial movement of the jaws *D* is obviously accomplished by the revolution of the scroll ring which meshes with teeth cut on the under sides of the jaws in the regular way. The body of the chuck is bored out at *C* to fit the locating plug *B* which is ground to the size of the hole in the work. The jaws *D* are bored out at *E* to the diameter of the gear blank at this point, and they are recessed at *F* as a precaution against chips or dirt. It will be seen that while the locating is done by the plug, the jaws grip the work firmly and act as drivers without any tendency to throw any strain on the plug. The scroll, having a floating action, takes care of any inaccuracies. This arrangement is rather unusual, but its action was very satisfactory in this instance.

Split Chuck for a Bearing Sleeve. — Fig. 15 shows a bearing sleeve which has been turned at *B* and *D* and faced at *C* in a previous operation. The fixture shown was made for facing

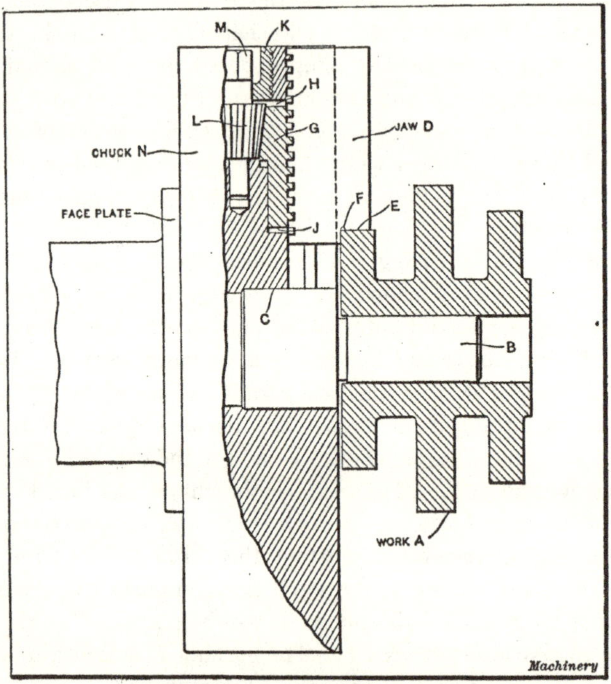

Fig. 14. Holding Device employing a Three-jaw Chuck with a Floating Scroll

the other end and boring the two surfaces *S*, but it was not entirely satisfactory, due partly to a tendency to remain closed after the clamping ring was unscrewed. A lack of spring in the chuck body was the cause of this; as it was made of machine steel, a great deal of elasticity could not be expected. Another bad feature was the position of the thread *N* which was so

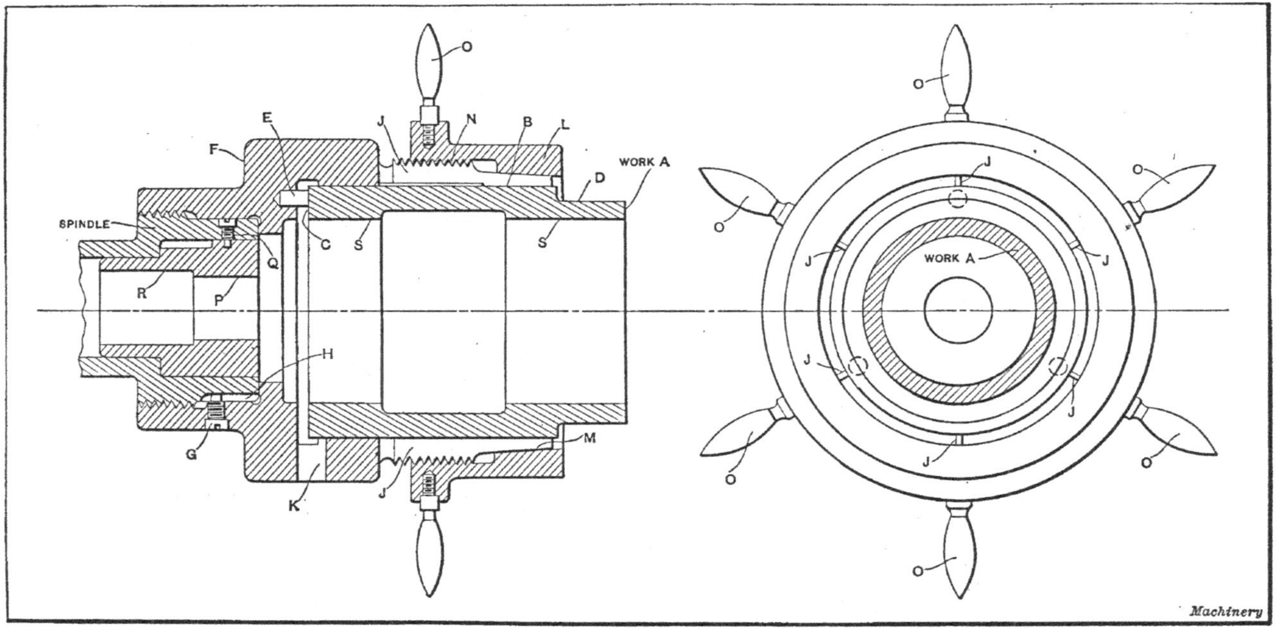

Fig. 15. Split Chuck used for a Bearing Sleeve

situated that considerable trouble was experienced in keeping it free from chips and dirt. This example is given to illustrate some of the faulty points in its construction, so that "what-not-to-do" may be clearly seen. The body of the chuck *F* was made of machine steel and was pack-hardened to obtain as much spring temper as possible. This body was screwed onto the spindle and held in place by the teat-screw *G*, which enters a slot *H* in the end of the spindle. Three pins *E* act as longitudinal stops for the work.

The inside of the chuck body is bored out to fit the work. It is threaded on the outside at *N* and tapered at *M*. Six saw cuts *J* are equally spaced around the periphery to allow for expansion and contraction. An operating collar *L* is threaded and tapered to fit the body, and six handles *O* screwed into place for operating purposes. These handles were afterward removed by the operator (as a precaution against being injured by them) and a piece of drill rod used in the holes in place of the handles. A tool steel bushing was fitted to the spindle and held in place by screw *Q*. This bushing was relieved at *R* and ground to a running fit for the boring-bar pilots at *P*. Three holes *K* were drilled in the body so that accumulated chips and dirt could be readily brushed out. This entire chuck exhibits a lack of forethought in its design and forms an example of a poorly constructed device. The recognition of its shortcomings has been of great assistance in subsequent designs.

Contracting Pin Chuck for a Small Flywheel Casting. — A somewhat expensive device is shown in Fig. 16 for handling the small casting *A* on a horizontal turret lathe. It will be noted that the arrangement of the chuck is such that the weight is kept well back near the spindle bearing, so that excessive overhang is avoided. The body *C* is made of cast iron and is screwed onto the spindle *D*, where it is secured by the teat-screw *E*, access to which is obtained through the hole *F*. Three pins *B* serve as longitudinal stops for the work, and these are so placed that they can be very easily kept free from dirt and chips. The three jaws *K* are cylindrical, and are forced outward by the coil springs *L*. The teat-screws *J* enter slots in the jaws and

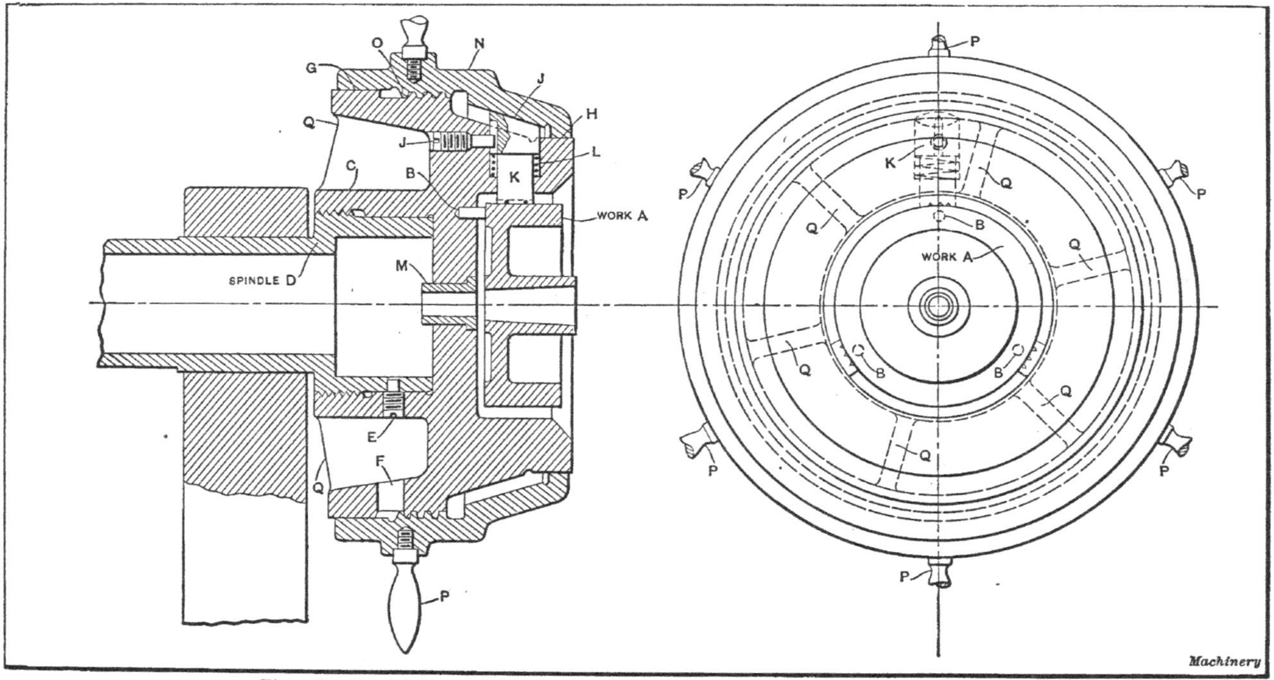

Fig. 16. Special Type of Chuck with Clamping Pins for a Small Flywheel Casting

prevent them from turning. The jaws are hardened, ground and lapped cylindrically, after which they are assembled in the chuck body and ground on the taper and also at their bearing points on the work. The operating collar *N* is a steel casting which is ground to a nice running fit at *H* and *G*, and threaded with a coarse pitch Acme thread (double) at *O*. It is ground to a taper at *J* corresponding to the taper on the ends of the jaws.

It is well to note that the construction and accuracy of the chuck would have been materially improved by making a tool steel taper ring and inserting it at this point in order to minimize the wear and provide for means of adjustment. The handles *P* are a rather dangerous feature, but they permitted the operator to use both hands in tightening the jaws; this was of considerable advantage, as the power required took about all the strength he had. A more acute angle on the pins in their relation to the collar would have been of some assistance, but the thread diameter and frictional surfaces were large, so that ease of operation could not be readily attained. The bushing *M* acted as a guide for the tools used in generating the tapered hole. It should be noted that the threaded portion is made somewhat free and the fitting done at *H* and *G*. These closely fitting surfaces are also of assistance in keeping chips and dirt out of the mechanism.

Vertical Turret Lathe and Boring Mill Fixtures. — As the class of work which is usually handled on machines of this type is of a heavier variety, it naturally follows that the fixtures required are much more massive, and also that they must possess ample driving means, on account of the heavier cuts taken. As the fixtures lie flat on the table of the machine, the weight is of no particular importance, but rigidity and proper protection against chips must be carefully considered in the design. In addition to this it is well to remember that outlets for chips are also of importance, so that the latter can be brushed out. If some provision is not made for this, it will be necessary for the operator to scoop out the chips from time to time, and this is an unnecessary waste of time. Outlets can nearly always be provided so that the matter will not be troublesome.

Special Jaws for a Large Motor Gear. — The work *A* shown in Fig. 17 is a steel forging for an electric motor gear. Both

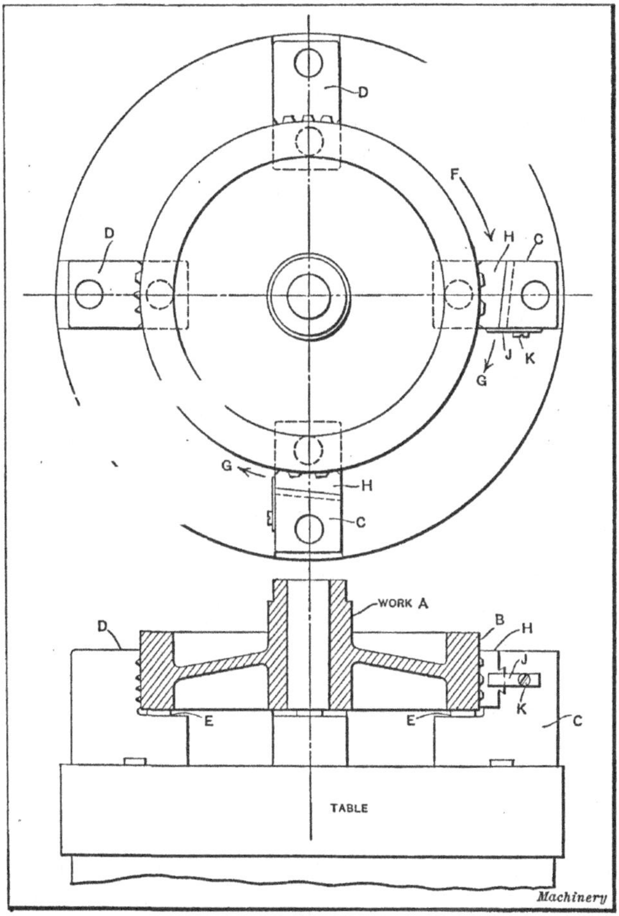

Fig. 17. Chuck for a Boring Mill having Special Secondary Jaws

the first and second settings of the pieces were accomplished on a vertical turret lathe; the operations were roughing cuts to within approximately $\frac{1}{8}$ inch of the finished sizes. The blanks weighed something over 350 pounds and were rough-finished all over in these two settings, about eighty pounds of metal being removed. The blanks were composed of alloy steel containing a high proportion of carbon and manganese, thus making them very tough. The speeds used were about twenty to twenty-five feet per minute, and the feed from $\frac{1}{16}$ to $\frac{1}{8}$ inch per revolution. It will be seen from this that chuck jaws of the regular type would be put to their utmost capacity to hold the work without chance of slipping. A few of the blanks were machined in the regular type of jaw and there was more or less slippage of the work when the heavier cuts were being taken.

The method shown was suggested by the author, but was not used on account of the few pieces to be handled at the time. The idea, however, is of some value. The two jaws *D* are of the regular type and are left set to form a sort of V-block arrangement for locating the work. The other two jaws *C* are of special design and are operated independently. Each of these jaws is fitted with a sub-jaw *H* sliding in a dove-tail slot in the main jaw. This slot is cut at an angle of 15 degrees from the tangent to the work, and the jaw is kept in its normal position by means of the flat spring *J* which bears against it. The screw *K* holds the spring in place. The jaws are all provided with raised buttons on which the work rests. It may be readily seen that the pressure of the cutting tools would tend to move the piece in the direction indicated by the arrow at *F*, in case any slippage were to occur. Assume, then, that this takes place during the progress of the work; a wedging action would immediately begin at the points where these special jaws are, and this would prevent further slipping.

Large Spring Collet Chuck having a Floating Action. — The flywheel casting shown at *A* in Fig. 18 represents a rather unusual condition, for the shoulder on the upper part of the hub is to be machined concentric with the tapered hole which has been machined in a previous setting. The rim, web and body of the hub are also to be machined in the second setting. A tool-steel, hardened and ground plug *E* fits the hole in the table and is a forced fit in the base F at *D*. The upper part of the

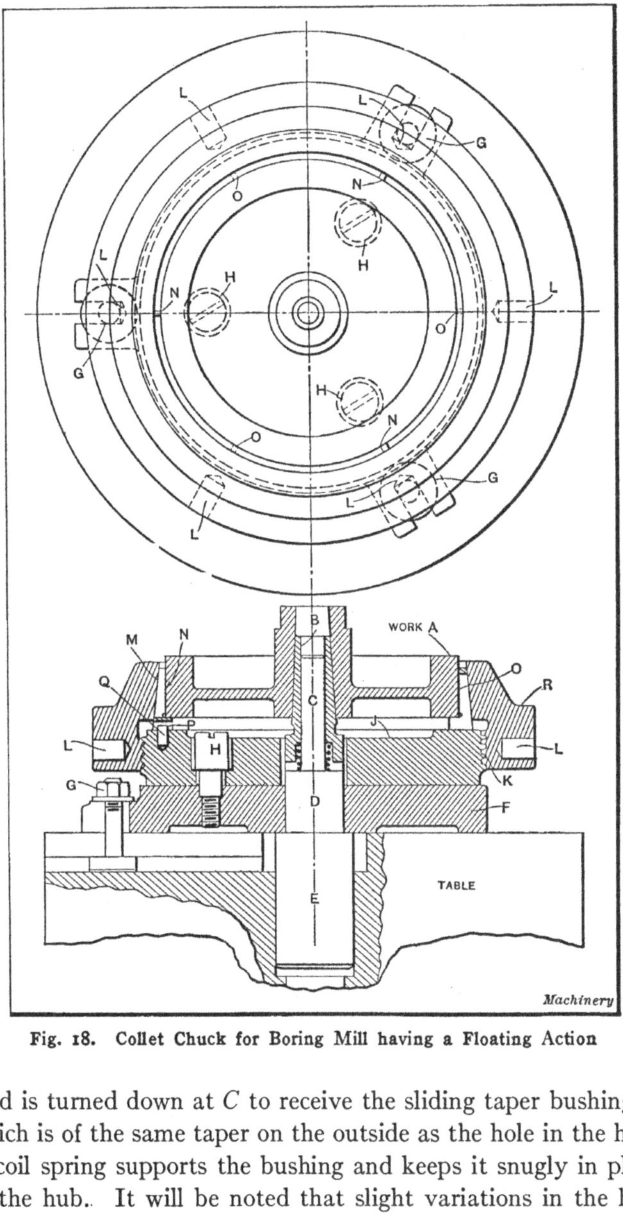

Fig. 18. Collet Chuck for Boring Mill having a Floating Action

stud is turned down at *C* to receive the sliding taper bushing *B* which is of the same taper on the outside as the hole in the hub. A coil spring supports the bushing and keeps it snugly in place in the hub.. It will be noted that slight variations in the hole do not affect the locating, as the sliding movement of the taper bushing equalizes them. The base *F* is of cast iron and is fastened down by the three bolts *G* in the table T-slots. The

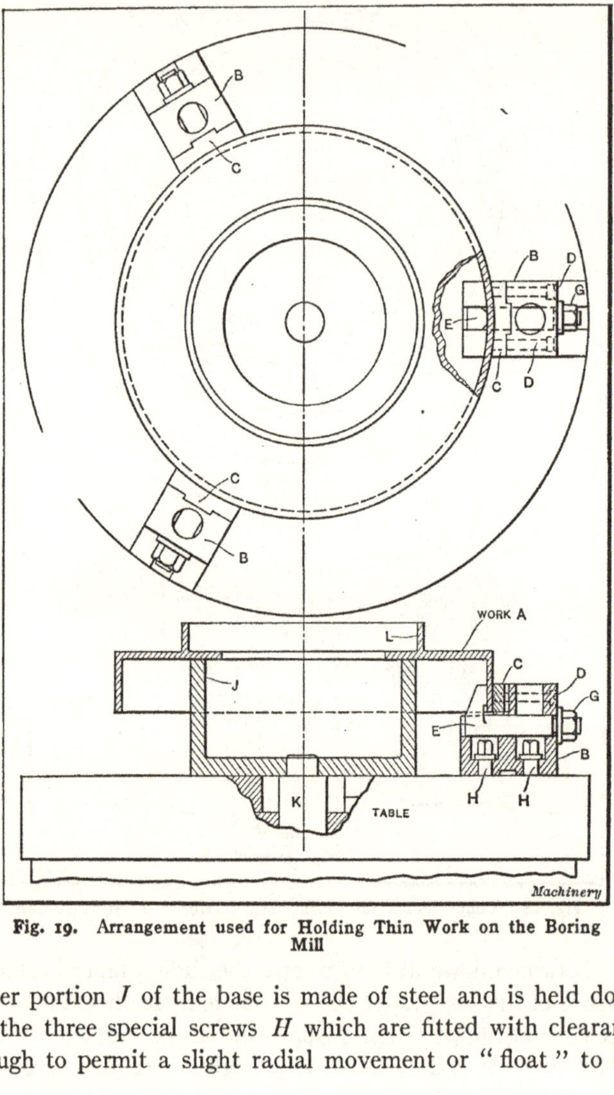

Fig. 19. Arrangement used for Holding Thin Work on the Boring Mill

upper portion *J* of the base is made of steel and is held down by the three special screws *H* which are fitted with clearance enough to permit a slight radial movement or " float " to the plate, and at the same time act as drivers. Acme threads, 4 per inch, are cut on base *J* to engage the operating collar *R*. A thin steel taper ring is slotted in six places *N* and *O*, and driving pins *Q* are provided which engage the slots *P* in the ring. The operating ring *R* is of cast iron, tapered at *M* to fit the spring ring. It is revolved by a drill rod handle set into any of the holes *L* in the periphery. In using the device, the work is dropped down on the spring tapered plug, which automatically adjusts itself to the hole and allows the rim of the flywheel to rest on the flange of the spring ring. It is then only necessary to revolve the operating ring, which clamps the work securely, the floating action taking care of slight inaccuracies.

Arrangement for Holding a Piece of Thin Work. — The thin steel casting shown at *A* in Fig. 19 was held in the first setting by the thin flange *L* in a set of hook-bolt jaws. During this setting the large diameter was machined both outside and inside, the inside of the web faced, and the hole through the web bored. In the second setting a cast-iron pot *J* is located on the table by means of the centering plug *K* which fits the center hole. This pot is used as a support for the work while facing and turning, and also locates the piece vertically. A set of jaws *B* is provided with hook-bolts *E* which are drawn up against the work on the inside by the nuts *G*. The soft steel sub-jaws are inserted in place and fastened by the screws *D*. These soft jaws are bored out to the correct size to fit the finished work and are brought up lightly against the casting before the hook-bolts are tightened. The screws *H* enter shoes which fit the T-slots in the table jaws. It will be noted that the soft jaws can be easily replaced, thus making the life of the main jaws almost unlimited.

Special Fixture for Holding Several Sizes of Bevel Gear Blanks. — The work *A* in Fig. 20 is one of ten blanks, ranging in size from 12 to 20 inches in diameter, and the fixture shown was made to handle all these sizes, by the simple expedient of moving the main jaws to an approximately correct position and inserting a set of soft jaws which are then bored to the exact size of the outside of the ring. The body *C* of the fixture is located centrally on the table by the plug *H* which fits the fixture at *J*. The four bolts *G* hold the casting down on the

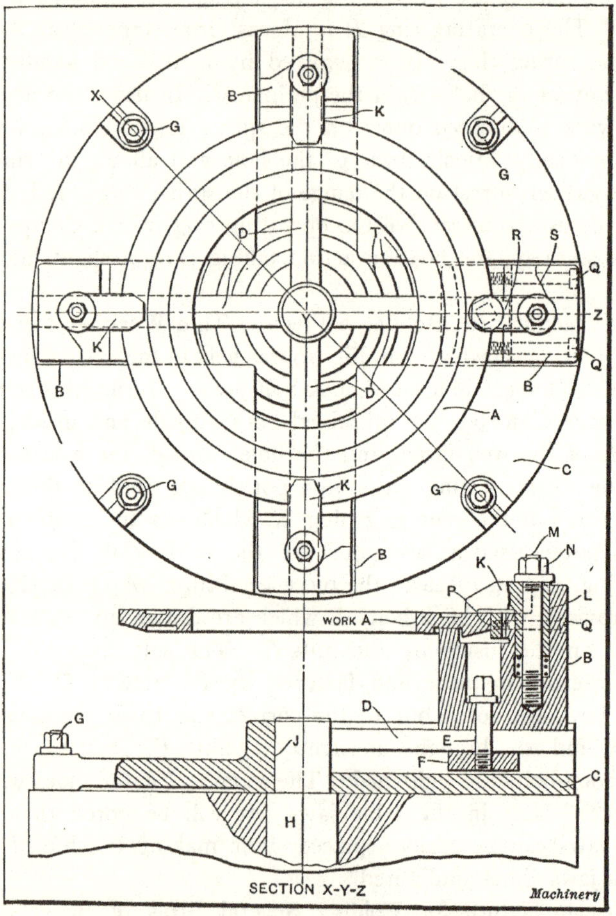

Fig. 20. Special Fixture for Holding a Number of Different Sizes of Bevel Gear Blanks on the Boring Mill

table. Four T-slots are cut at *D* and the main jaws are clamped in their proper positions by bolts *E* which enter the shoes *F* in these slots. Approximate locations are determined for the jaws by the radial lines *T* scored on the finished pads. Both the main and sub jaws are of machine steel, the former pack-hardened and the latter left soft. The main jaws *B* are tongued on their lower sides to fit the slots. The soft jaws *P* are made up in a long strip and then sawed up into separate pieces. They are tongued at *R* to fit the main jaws and jig-drilled for the screws *Q* which hold them in place. The hook-bolts *K* are of tool steel, hardened and drawn to a blue on the hook end, to lessen the chance for breakage. The body *L* fits the hole in the jaw and the coil spring supports it. The stud *M* is screwed tightly into the jaw and is threaded at its upper end to receive nut *N*, which operates the hook-bolt. Attention is called to the manner in which the hook-bolts are backed up by the jaw, thus greatly stiffening the clamping arrangement. The backing is partially cut away at *S* to allow the hook-bolt to swing clear of the work. This fixture gave excellent results.

CHAPTER VII

METHODS OF MACHINING THIN AND IRREGULAR WORK

Handling Thin Work. — In nearly every kind of manufacturing it becomes necessary to machine certain pieces of work, which are of a fragile nature, and which can neither be held nor machined by ordinary methods without considerable distortion. Sometimes the work is a casting of very thin section, while in other instances a forging is to be cut down until the walls are not over $\frac{1}{8}$ inch in thickness. The work may be of comparatively small diameter or it may be of large size, and it may also be either long or short. It may be cylindrical in form or of irregular shape, and may occasionally require the addition of holding lugs in order to hold it properly. In the machining of this class of work we shall consider two types of machines only — the horizontal and the vertical turret lathes. The horizontal type of machine would naturally be used for the smaller work, while the larger pieces can be more profitably handled on the vertical machine.

While work of this character must be very carefully held in order to avoid distortion, the problem of machining is also of great importance. It is quite possible to hold a piece of work so that it will not spring out of shape, and yet the machining operations may be very unsatisfactory due to vibration in the work itself. This causes chatter which nearly always ruins both the work and the tools used in machining. Vibration is more apt to be troublesome on long work than on the shorter pieces, because the torsion or twisting action induced by the pressure of the tools in removing stock is more apparent when this action takes place at a considerable distance from the points at which the work is supported and held. On short work the support is nearer the point at which the work is being done, and there is consequently less chance for the metal to twist under the pressure of the cut.

Influence of Feeds and Speeds. — In connection with the machining it is well to note that the speed and feed have a great influence on the vibration, but it is difficult to give a definite rule which will apply to all conditions. Methods of holding have a great influence on this matter, for it is obvious that the more securely the work is held, the greater may be the feed and speed, other things being equal. Speaking generally, less vibration results when slow speeds are used with fairly coarse feeds, but this is not always the case. Sometimes it is possible to run the work at a fairly good speed, using a fine feed, while at other times a procedure of this kind will cause chatter which can be heard all over the factory. An increase of one step in the feed or a decrease of one step in the speed will frequently stop the trouble, but there are occasional instances which require considerable experimenting before the desired result is obtained.

When outside and inside cutting are going on at the same time, it is a good plan to arrange the boring and turning tools so that they are working opposite each other; that is, the tools should be working one against the other with nothing but the wall of the casting between them. If used in this way there is less chance of springing the work, and also less chance of vibration. In handling thin work, special care must be taken to see that all bearings, gibs, etc., are set up snugly, so that no chatter will be caused in the machine itself. The tool-holders and the tools themselves must also be as solid as possible.

Important Points relating to the Handling of Thin Work. — A few points in connection with the handling of work of this character may be of interest and are given in the following paragraphs:

1. The method of holding should be very carefully thought out with a view to rigidity and freedom from distortion of the work itself. It may be advisable in some cases to have supplementary lugs or pads added to the pattern in order to facilitate chucking. When this is necessary it is well to consult with the pattern-maker, so that unnecessary expense in the pattern will be avoided.

2. The overhang from the spindle should be as short as possible, so that vibration will not be caused by an excessive weight revolving without support at its outer end.

3. The rapidity of operation is an important point if a great many pieces are to be machined. Clamps should be easily accessible and conveniently operated. It is advisable to make all nuts of the same size so that one wrench can be used for all. It is also desirable to make them of such a size that a standard machine wrench can be used.

4. The locating surfaces, if the work is irregular, should be carefully selected, so that inequalities of the casting will be equalized as far as possible. Care should be taken so that no locating points come where the casting is gated or where letters or numbers appear. Neither should they come where the pattern is parted.

5. The tooling should be very carefully studied both from the viewpoint of upkeep and of rigidity, so that every possible precaution may be taken to prevent chatter.

6. The cost of equipment should be considered on the basis of the number of pieces to be machined. Obviously, it does not pay to design very elaborate equipment for work which will not be machined in large quantities.

7. The selection of a machine best adapted to the work is an important point, and should be decided upon before anything is done in the line of equipment.

Machining a Thin Steel Casting on a Horizontal Turret Lathe. — The work shown at *A* in Fig. 1 is a steel collar which is to be finished all over on a horizontal turret lathe. When in the rough, the thickness of the wall of the casting is $\frac{5}{16}$ inch, and it is to be finished to $\frac{3}{16}$ inch. As the work is so thin that the action of the chuck jaws would tend to crush it out of shape, if they were used in the ordinary manner, the driving lug *C* was added to the casting. In this way much less pressure is required on the jaws than would be the case if friction only were used for driving. It will be noted that the jaws *B* have a narrow shoulder against which the work rests. The body *G* of the boring tool is of cast iron and is fastened to the turret by the angular gib J and the screws H. A pilot E of tool steel is hardened and ground to fit the bushing D, which is forced into the chuck body. The tool F is set in a vertical plane in the holder in order to minimize errors in indexing the

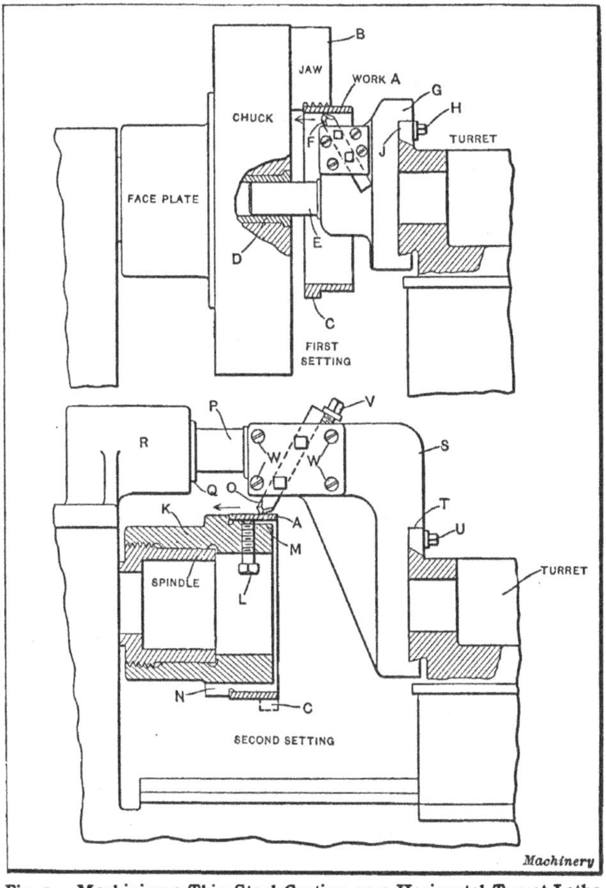

Fig. 1. Machining a Thin Steel Casting on a Horizontal Turret Lathe

turret. Two holders of this kind are used for the work, and the cut-off slide tools face the end of the collar. While the piece is being rough-bored, the jaws are set up firmly on the work, but care is used so that too much pressure is not applied. Before the finish-boring takes place the jaws are loosened up to take care of the spring of the casting.

In the second setting (shown in the lower part of the illustration), the work *A* is pushed onto the arbor *K* which is screwed onto the end of the spindle; the cup-pointed driving screw *L* is set up against the finished inside surface. A shallow groove *M* is provided so that the burr thrown up by this screw will not cause trouble when removing the work. A moment's work with a scraper takes out the slight roughness after the work has been removed from the arbor. The grooves *N* on the periphery of the arbor facilitate the removal when the casting has been machined. The body of the turning tool *S* is fastened to the turret by gib *T* and screws *U*, and it has a pilot of steel at its forward end *P*. This pilot enters a bushing *Q* in the spindle cap bracket *R*, thus greatly assisting in the prevention of chatter. A steel plate is fastened to the body by four screws *W*, and supports cutting tool *O*. A refinement will be noted in the backing-up screw *V* which permits fine adjustments to be made. It was found advisable to cut off the lug *C* with a hacksaw previous to this setting, as the interrupted cut produced when attempting to machine it off with the turning tool tended to twist the work on the arbor.

Machining a Sliding Sleeve.—The sliding sleeve shown at *A* in Fig. 2 is of cast iron, ⅛ inch thick when finished. Its outside diameter is important and its over-all length also. It is likewise essential that the two ends should be parallel, and square with the outside. The casting is made up in the form of a pot with three beveled lugs spaced 120 degrees apart. The end of the pot is rough-ground on a surface grinder, and three holes *C* are jig-drilled to act as drivers and locaters. This work was done previous to the machining of the piece on the turret lathe. The fixture body *E* is then screwed to the spindle of the machine and is furnished with driving and locating pins which enter the drilled holes. The hook-bolts *D* are beveled to the same angle as the lugs and extend back through the body so that they may be tightened from the rear, thus leaving the front of the fixture free from projections and permitting the turning tool to work without interference. Attention is called to the manner in which the hook-bolts are backed up by the boss, which is cut away at one side so that the bolt can be turned out of the way when assembling or disassembling the work.

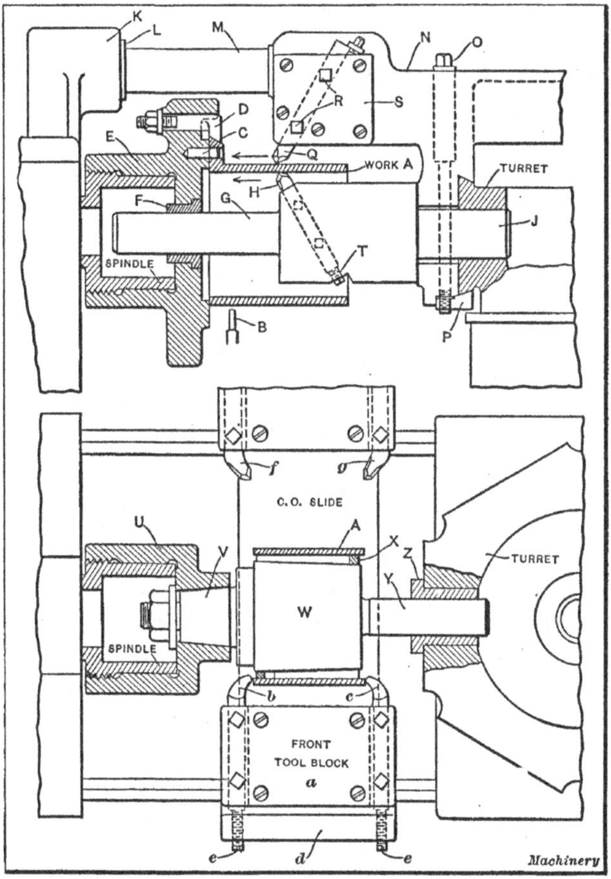

Fig. 2. Machining a Sliding Sleeve in a Turret Lathe

The body of the fixture is provided with a bushing F which acts as a guide for the pilot of the boring-bar *G*. This bar is firmly secured in the turret at *J*, the regular turret binder being used to hold it. The tool *H* is adjustable for diameters through the backing-up screw *T*. The outside turning is accomplished by means of double-end turning tool *N* which is fastened to the turret faces by the gibs *P* on the lower dovetailed faces, these being secured by the long special screws *O* which pass down through the body of the tool. A steel pilot *M* enters the bushing *L* in the spindle cap bracket *K*. A steel plate *S* is mounted on the body, and tool *Q* is secured in place by two screws *R* which pass through the plate. A collar-head screw is provided by means of which fine adjustments can be readily made. Attention is called to the manner in which the turning and boring tools are arranged so that they are working opposite each other, thus tending to keep the work nearer to size than would be the case if they were set to cut on different portions of the casting. After the work has been finished, the parting tool *B* cuts off the end of the sleeve slightly longer than required, so that a finish allowance is made for the second setting.

The second setting is shown in the lower part of the illustration and consists of roughing and finish-facing both ends of the sleeve to the prescribed size. For this setting the work *A* is held on the arbor *W*, the expanding sleeve *X* being forced up on the taper in the usual manner. A nose-piece *U* is screwed to the spindle and receives the tapered end of arbor *V*, while the other end *Y* is hardened and ground to a nice running fit in bushing *Z*, located in the turret. A special tool-block *a* is mounted on the front of the cut-off slide and holds the two rough-facing tools *b* and *c*. The tools enter slots in the block and are held in place by the set-screws shown. A ledge is cast at *d* on the rear of the block to allow the use of adjusting screws *e*. The finishing tools *f* and *g* are mounted in a special block on the rear of the cut-off slide in the same manner as those in front.

Machining a **Thin Flanged Collar** on the **Turntable Lathe.**—A turntable lathe made by the Pratt & Whitney Co. is used to machine the collar shown in Fig. 3, the cross-sliding turret of this machine being used to perform the facing operation on the flange and the cutting of the recess. The work *A* is held in the special jaws *B* of the three-jawed chuck, that portion of the jaw marked *D* being brought up lightly against the inside of the thin end of the work and the set-screw *C* tightened. The forward end of the jaw is used as a support for the flange and also serves to locate the work longitudinally. The boring-bar

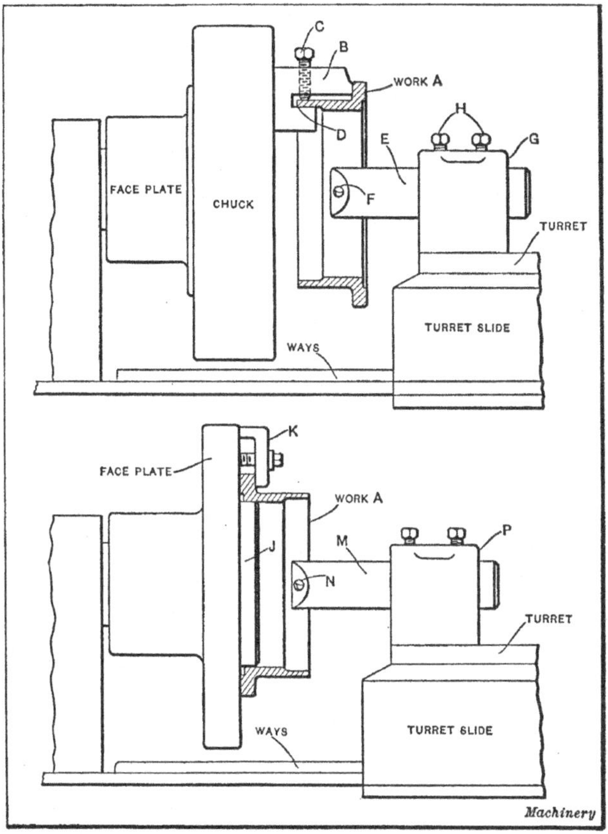

Fig. 3. Thin Flanged Collar Machined on a Pratt & Whitney Turntable Lathe

E holds the tool *F* which does the boring, the turret being set off center a sufficient amount to bore the required diameter. The tool-holder *G* is one of the regular type commonly purchased with the machine. The two set-screws *H* are used to force down a beveled shoe which, in turn, serves to contract a split bushing which grips the bar. In connection with this equipment attention is called to the fact that the jaws are used principally for centering the work, being brought up very lightly on the inside of the work so that they have no tendency to distort the casting. The set-screws are then set up tightly on the outside, and as the piece is backed up by the jaws on the inside, there is no danger of springing. A metal-to-metal contact is obtained in this way and any tendency toward vibration is avoided.

The second setting of the work is shown in the lower part of the illustration. For this setting a faceplate is used with a locating plug *J* which fits the previously bored interior, the flange face being brought up against the finished face of the plate. Three clamps *K* secure the work. The boring-bar *M* holds tool *N* for boring the thin portion of the work, the turret being set off center for this purpose as in the first setting. The bar is held in the same type of holder *P* as that shown above.

Machining a Steel Motor Casting. — The work shown at *A* in Fig. 4 is a steel motor casting of somewhat fragile construction at its inner end. This is another instance in which the turntable lathe is used. It will be noted that the casting is cored out at its inner end, so that there are three elongated openings at this point. In order to clamp the work without distortion, a steel ring *C* is provided with three set-screws *D*. This ring is placed in position in the casting before it is gripped by the jaws, the set-screws being set up with a moderate pressure against the inside of the walls. The casting is then placed in the chuck in such a way that these set-screws come opposite to the jaws. The jaws *B* may now be set up firmly without danger of crushing the work, the pressure being taken by the screws. The double tool-holder *K* is fastened to the turret face, and the set-screws *L* and *M* clamp the bars in position. These bars extend across the turret and the other ends are used for the finishing tools. The turning bar *J* is provided with a tool *H* which goes through the bar at an angle so that the tool is slightly in advance of the bar, thus permitting work to be done close up to a shoulder. A backing-up screw is provided for fine adjustments. The boring-bar *F* is arranged in the same manner, tool *E* being backed up by screw *G*.

The second setting is shown in the lower part of the same figure, the work *A* being held on a special arbor *O* which is screwed onto the end of the spindle. The finished end of the work is brought up against the shoulder on the arbor, the work being held on the portion *P* which fits the inside of the casting.

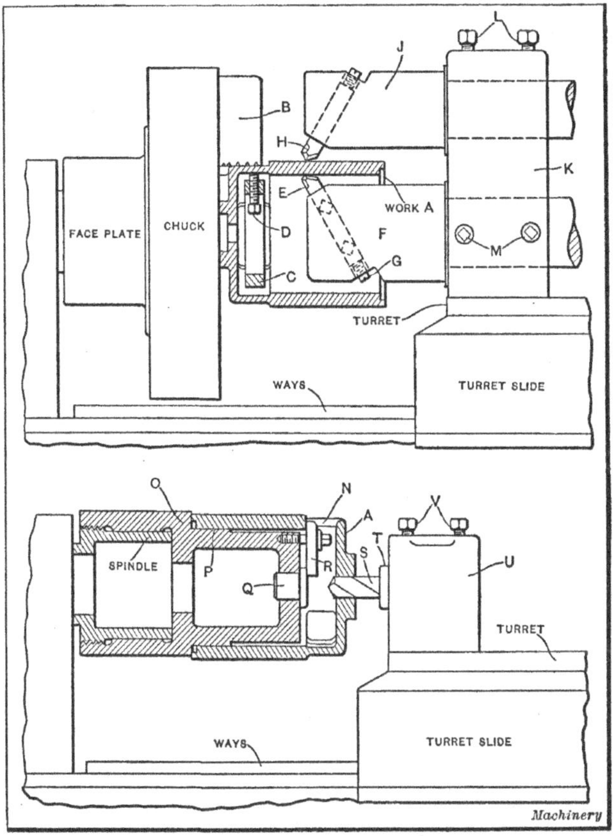

Fig. 4. Steel Casting of Fragile Construction Machined on Turntable Lathe

The clamps *R* enter the cored holes in the casting *N*, and the steel stud *Q* forms a support for their inner ends. The tool-holder *U* is of the same type as previously described, the bushing *T* being compressed by two screws *V*, so that it holds the drill *S*. Other tools used in this setting are of a simple nature and

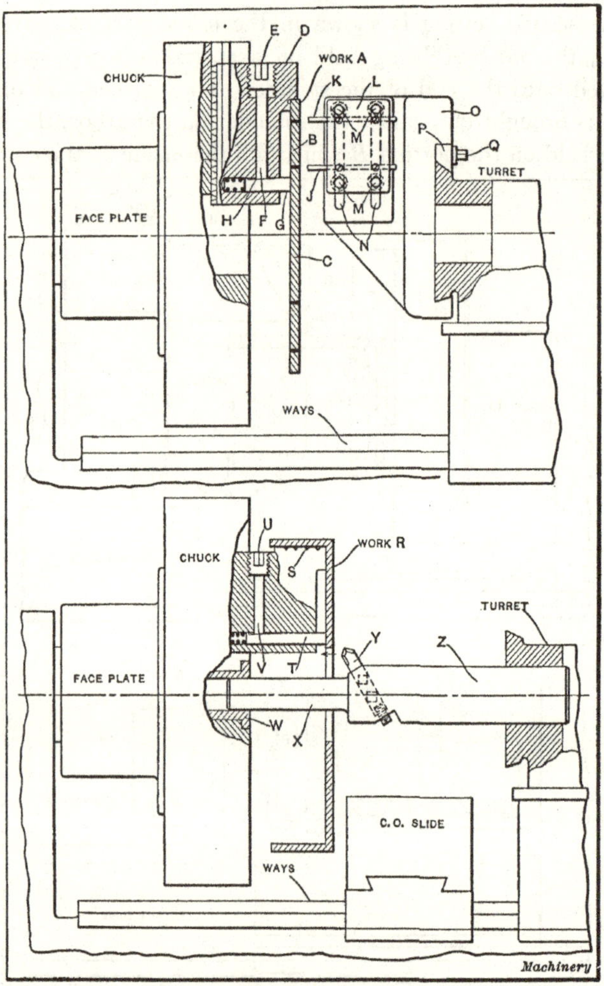

Fig. 5. Cutting out a Thin Sheet Steel Collar and Turning, Boring and Facing a Thin Steel Drum

need not be described. The facing is obviously done by the transverse movement of the turret.

Cutting Out a Thin Sheet Steel Collar. — The illustration in the upper part of Fig. 5 shows a piece of hexagonal plate at *A* from which the thrust washer *B* is to be cut. In this work the center part of the blank *C* is saved and used for other work, but the outside portion is scrapped. The outside of the blank is held in the special jaws *D*, each of which is provided with a spring plunger *G*, a binding plug F and a hollow set-screw *E*. One side of the plunger is flatted off with a slight back taper as shown at *H*, so that any tendency to push back is offset by the wedging action of this taper.

When the device is in operation, the plungers are first pushed down out of the way, and the jaws tightened, after which the plungers are released until they come in contact with the plate. The screws are now tightened and the work is ready for machining. A special tool-holder *O* is secured to the turret by angular gib *P* and screws *Q*. A steel tool-block *L* is mounted on a finished pad and is held in any desired position by the four screws *M*, which pass through the block and the body of the holder and are secured on the other side by nuts and washers. Vertical movement of the block is permitted by the elongated slots *N*, thereby allowing a number of different diameters of washers to be cut with the same tools. The tools *J* and *K* are ground on the sides to fit the slots in the block, and it should be noted that the lower of the tools is set forward slightly in advance of the other so that the center piece *C* will be cut out before the separation of the outer ring occurs.

Turning, Boring and Facing a Thin Steel Drum. — The lower illustration in Fig. 5 shows another piece of work in which the same type of jaw with spring plungers is used to support the casting. In this case the jaws are so made that they grip the inside of the shell at *S*, while the plungers *T* come up against the web and assist in supporting it against the pressure of the cut. The same construction is used in the binding plug *V* and the hollow set-screw *U* as in the upper illustration. The facing of the web is performed by the cut-off slide tools while the boring-bar *Z* is cutting out the central portion with the tool *Y*. The forward end of this bar is hardened and ground at *X* to a running fit in the chuck bushing *W*. This piece of work was completed in another setting by holding it in soft jaws which extended in to form a support for the web.

Machining a Thin Ring Pot on the Vertical Turret Lathe.— We now come to work of a larger size which can be handled to good advantage on the vertical turret lathe. On this type of machine the work stands vertically, so that there is no overhang of the spindle to contend with as on the horizontal type of machine. Aside from this, the conditions are similar in

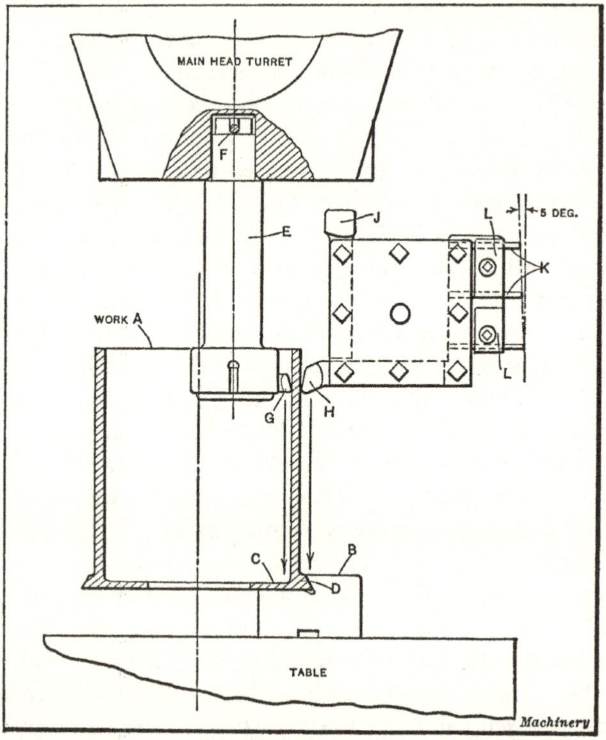

Fig. 6. Machining a Thin Casting to be cut up into Rings on the Vertical Turret Lathe

both types. As the tools hang vertically, excessive overhang from the turret is not so apt to produce chatter as in the horizontal type of machine.

Fig. 6 shows a cast-iron ring pot from which four thin rings are to be machined. The pot casting *A* has a beveled flange around its lower end and is reinforced on the inside by web *C*.

The jaws *B* are beveled to the same angle as the flange at *D* and, therefore, have a tendency to draw the work down as well as to center it. The inner web tends to prevent crushing the pot when pressure is applied to the jaws. The bar *E* is a combination boring and reaming bar (as manufactured by the Bullard Machine Tool Co.), and is provided with slip-cutters and floating reaming cutters. (This type of bar was fully described in Chapter II.) In this particular case the reaming cutters are not used, but roughing and finishing tools as shown at *G* are used in boring the casting. The shank of the bar fits the turret hole and is prevented from turning by pin *F*. The side-head turret holds roughing tool *H* and finishing tool *J*, these tools being used to turn the exterior of the pot.

It should be noted that the boring and turning tools are working opposite each other, in the manner previously described. A special supplementary tool-block is fastened to the opposite side of the side-head turret and carries the three tools *K* which cut off the rings. These tools are ground on the sides so that they fit the slots in the tool-block, and are held in place by straps tightened by screws *L*. Attention is called to the fact that these tools are set so that a line passing through their cutting edges forms an angle of 5 degrees with the perpendicular. This is done so that the rings will be cut clean without ragged and broken edges, as only one ring is cut off at a time.

The work *A* shown in Fig. 7 is a steel casting which is to be bored on the two internal annular pads, faced on the two ends, and turned on the outside. The casting is centered by the lower part of the special jaws *K*, these being brought up lightly against the casting so that they do not distort it. The screw dogs *G* on the inside of the work are then tightened until the pointed screws sink into the casting, thus holding the lower end firmly. The pointed set-screws *L* in the upper part of the jaws are now brought up against the work and tightened, after which the piece is ready for the machining operations. For boring the inside pads a special bar of the same type as that shown in Fig. 6 is used, the bar *D* being extra long. The shank *B* fits the turret and is driven by the pin *C*. A roughing and finishing slip-cutter *E* does the boring, while tool F in the side-head turret faces the end and turns part of the periphery.

The second setting is shown in Fig. 8, but between the first and second turret lathe operations a slot *M* is milled across the lower end of the casting to assist in driving. The work is

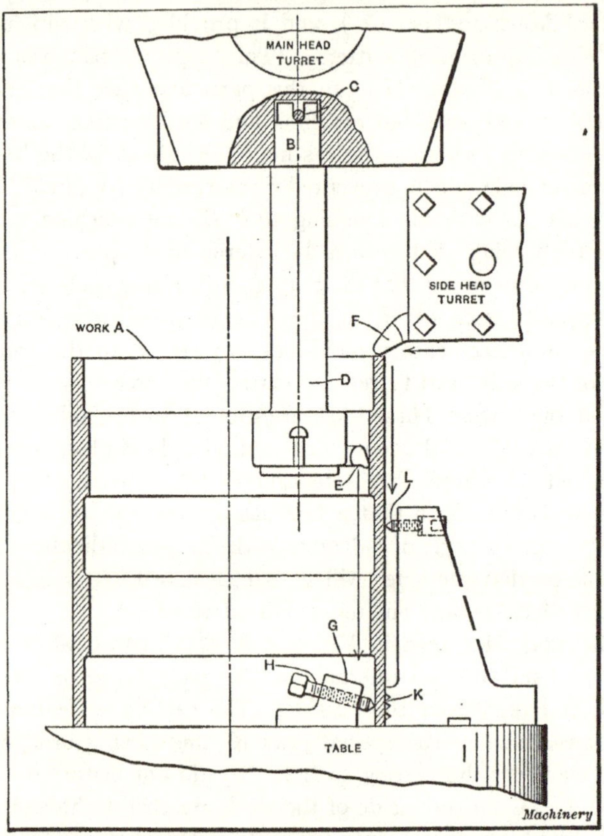

Fig. 7. Boring, Facing and Turning a Thin Casting in a Vertical Turret Lathe

held during the second setting on the special locating fixture *B*, which is made of cast iron and bolted to the table by the three bolts *O*, T-shaped at their lower ends and fitting slots in the table. The centering plug *K* is a drive fit in the fixture body and is turned down at *L* to fit the center hole in the table. A steel strip *N* is fastened to one side of the body and its upper end acts as a driver in the slot *M*. The lower part of the outside of the body is turned at *C* to a nice fit in the casting.

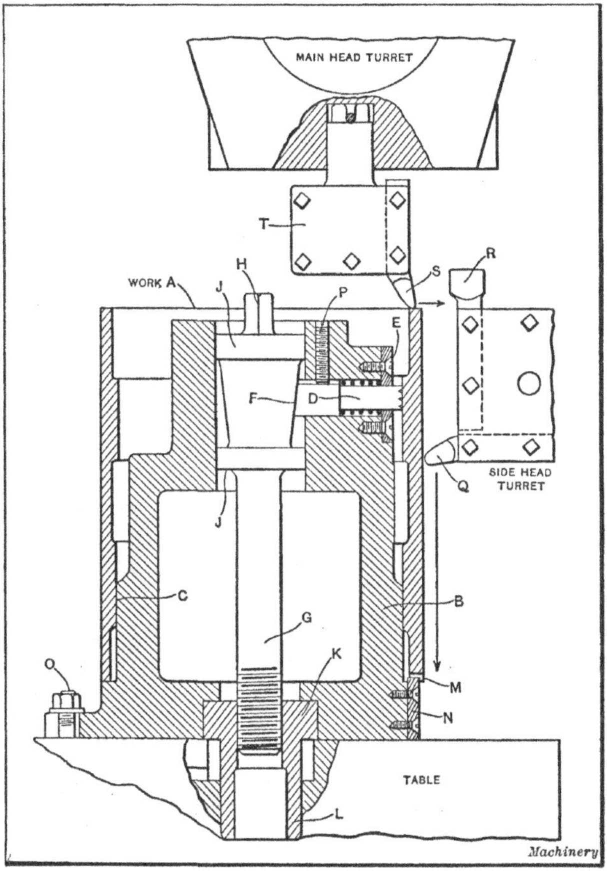

Fig. 8. Second Setting for Machining Casting shown in Fig. 7

The upper portion is firmly held by three expanding pins *D* which are flatted on one side where the screws *P* bear against them to prevent turning. The pins are shouldered for the coil springs and are beveled at their inner ends to fit the angle F on the operating rod. Three steel plates *E* are let into the casting to act as cover plates and hold the springs in place. The operating rod *G* is of tool steel with its lower end threaded with an Acme thread in bushing *K*. Two narrow rings on the upper end *J* are made cylindrical and a square portion is milled to receive a wrench by means of which the mechanism is operated. It will be seen that the purpose of the pins is to steady the upper portion of the casting and prevent torsion which might

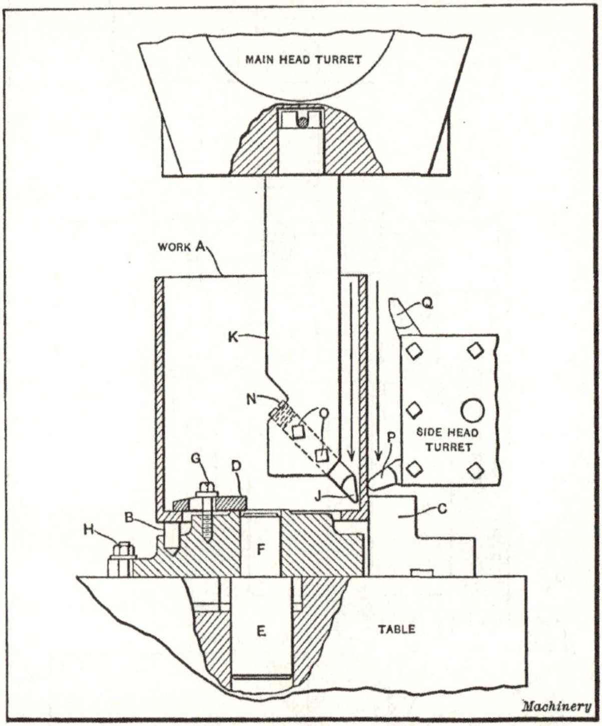

Fig. 9. A Thin Cast-iron Shell finished all over in a Vertical Turret Lathe

result under the pressure of the cut. Excessive pressure is not necessary on these pins, as a great amount of the driving is done by the block *N* at the lower end. The tool *S* is used for facing the end of the casting and is held in one of the regular tool-holders *T*, the shank of which fits the turret hole and is driven by a pin. The side-head holds two tools *Q* and *R* which are used for roughing and finishing the periphery.

Machining a Thin Cast-iron Shell. — The work shown at *A* in Fig. 9 is of cast iron and is finished all over, the side walls being only $\frac{3}{16}$ inch thick when finished. A cast-iron base is held down on the table of the machine by the three bolts *H*, which are T-shaped at their lower ends and fit the table T-slots. A steel plug is forced into the base at *F* and fits the center hole in the table at *E*, thereby locating it centrally. Three hardened steel pins *B* are inserted in the base and act as supporting points for the casting. The three clamps *D* are operated by screws *G* and are slotted so that they can be pulled back, out of the way, when inserting or removing the work. In setting up, the jaws *C* are used to center the work, but they are not set up very tightly for fear of buckling the casting. Two special bars *K* are used for the roughing and finishing boring; tools *J* are put in at an angle of 45 degrees so that there will be no interference with the clamps when near the bottom of the pot. Two screws *O* clamp the tools in place, and a backing-up screw *N* permits fine adjustments to be made. The tools *P* and *Q* in the side-head are used for roughing and finishing, and when in use are kept directly opposite the boring tools.

The second setting of the same casting is shown in Fig. 10, this operation consisting of boring the hole through the flange and undercutting it, and also approximately matching up the unfinished portion of the inside with that previously machined. The fixture body *B* is made of cast iron and is bolted to the table in the usual manner by the three bolts *D*. A locating stud *E* is forced into the body and fits the center hole in the table at *F*. The lower part of the body is turned at *C* to fit the inside of the bored shell. A shoulder is provided against which the finished end of the work is located, and a groove *Q* is cut so that trouble will not be caused by chips and dirt. A steel ring *G* is fastened to the upper end of the body by three screws *H* in such a way that there is clearance around the body of the screw; thus a floating or equalizing action is permitted. Three steel shoes *K* are let into slots in the ring and are controlled by screws *J* which pass through the ring.

It will be seen that this arrangement permits the screws to

be tightened sufficiently to perform their function without danger of forcing the work out of true, due to unequal pressure on the screws, as the floating ring equalizes the strain. The boring-bar *L* used for the work is of the same general type as previously described, slip-cutters being provided for both boring

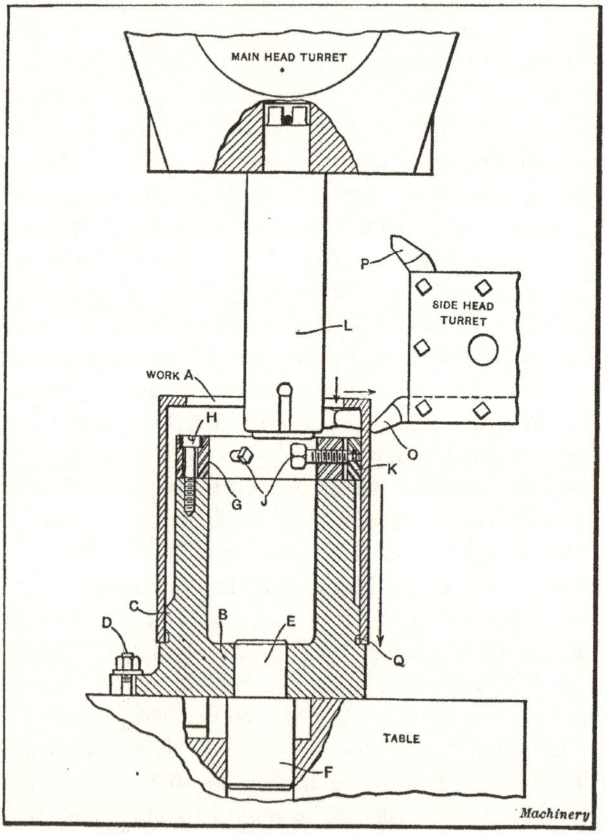

Fig. 10. Illustration showing Second Setting of Same Casting as shown in Fig. 9

and undercutting the flange. The facing of the flange is accomplished by tools carried in a regular tool-holder as shown in Fig. 8, while the outside turning is done by the roughing and finishing tools *O* and *P* in the side-head turret.

Machining a Steel Sprocket. — The work shown in Fig. 11 is rather out of the ordinary. The piece when completed is a steel sprocket, such as used on automobile trucks. Two settings are required to complete the work, but the first one of these is the only one in which we are interested. The pieces from which the sprockets are machined are octagonal in shape and are of rolled steel with a large hole punched in the center.

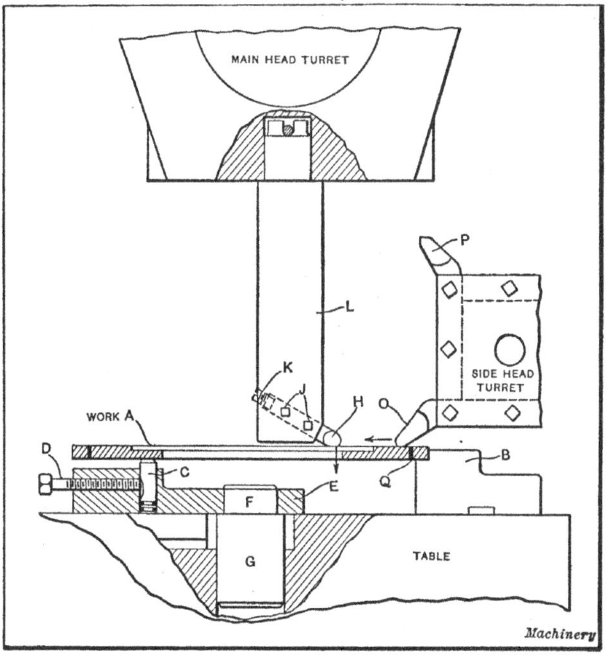

Fig. 11. Machining Steel Sprocket for Automobile Trucks

The work is roughly centered and gripped by the outside in the special jaws *B*. A cast-iron spider casting *E* is centered on the table by stud *G*, which fits the center hole in the table and the upper end F of which is a drive fit in the spider body. Three spring plungers *C* support the inner part of the plate, and are locked in their positions by set-screws *D*. In operation, these plungers are pushed down, out of the way, until the jaws have been tightened, after which they are released and the set-screws tightened. The tooling for this piece of work is not out of the ordinary, bar *L* being similar to the one shown in Fig. 9, except that tool *H* is put in at an angle of 30 degrees instead of 45. The screws *J* hold the tool, and the backing-up screw *K* permits of fine adjustments. The tools *O* and *P* in the side-head

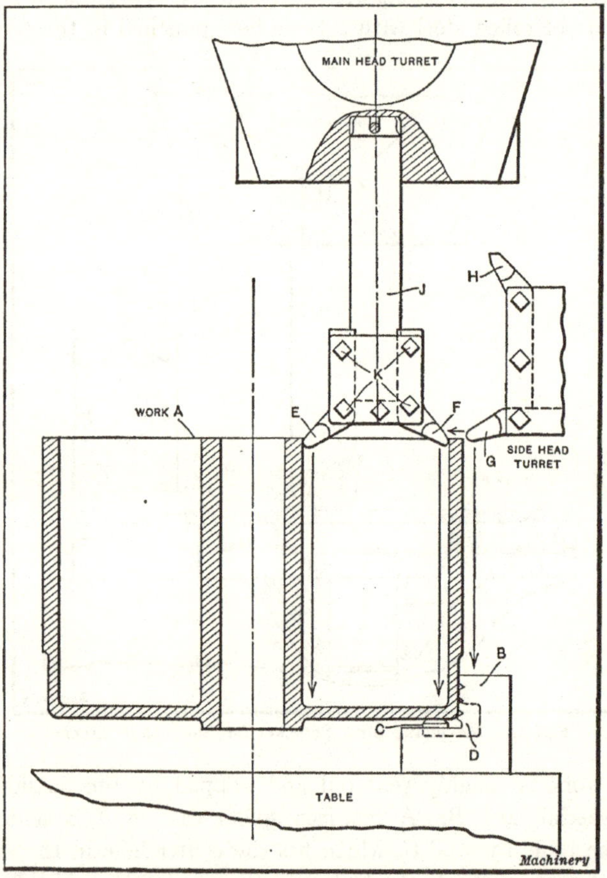

Fig. 12. Another Thin Casting Machined in the Vertical Turret Lathe

are used for roughing and finish-facing. The recess is cut by a tool in the main head, while the final operation of cutting out the work at *Q* is accomplished by a parting tool in the main head.

Machining a Large Pot Casting. — The work shown at *A* in Fig. 12 is of cast iron and is machined completely on the inside while the outside is turned as far as the finished pad extends. The work is held by the outside in the standard jaws *B*, supporting the bottom of the pot on the raising buttons *C*. A lug *D* on the outside of the pot acts as a driver against the side of the jaw. A long special tool-holder *J* holds the two tools *E* and *F*, these being held in place by set-screws *K*. These two tools turn the hub and bore the inside of the pot. At the same time that this operation takes place on the inside, tool *G* in the side-head turret is turning the outside diameter. The cutting points of all three tools are in a line. Another long tool-holder holds the tools for the inside finishing while tool *H* in the side-head completes the outside work.

Chucking Methods for Irregular Work. — Castings or forgings of irregular form usually require some method of holding (for the purpose of machining) other than chuck jaws, although there are many instances when the latter can be used to advantage. The form and size of the work have a great influence on the method of handling, and the accuracy required is also an important factor. The work itself may be in the rough state without any machining previous to the chucking operation or it may have been milled, planed or drilled. In the former case it is necessary to design a method of holding suited to the rough piece, and care must be exercised in selecting the surfaces best adapted for locating the work. It may even be necessary to have additional lugs or bosses cast or forged on the piece in order to facilitate holding it in position on the fixture.

If the walls are thin, care must be used also in the method of clamping so that no distortion will take place. In case a previous operation of milling or planing has taken place, it is essential that the fixture be so designed that this surface be used for locating in order that accurate work may be assured. There are occasional instances when both milling and drilling operations have preceded the chucking. This may possibly complicate matters somewhat or it may simplify them depending on the conditions. Sometimes a series of holes has been drilled in a flange which has been surface milled, and in a case of this sort the holes may prove useful for clamping purposes. When a case is encountered with a milled surface and an angular hole, or some other condition of a similar nature, there may be more difficulty in designing the fixture.

The type of machine upon which the work is to be done is also a factor which largely influences the design, and this matter should be decided positively before any attempt is made to draw up the device. The types of machines for which fixtures of this sort are most frequently designed are the horizontal and vertical turret lathes and the vertical boring mill. The engine lathe is also occasionally fitted with an arrangement for this kind of work, but as the construction of the fixtures is very similar to those used on the horizontal turret lathe it is not necessary to differentiate between the two in this chapter.

Important Points in Design. — It is obviously out of the question to attempt to cover or describe all the conditions which may be encountered in the chucking of irregular work, but the examples here given represent a variety such as may prove of considerable value in elementary design. Adaptations to suit various conditions will suggest themselves to the progressive designer. Attention is called to a few of the important points in the design of fixtures of this character.

1. Locating points or surfaces: These are very important and should be selected with care, having in mind any ribs, lugs or raised lettering on the casting, so that no trouble can be caused through faulty locating. Make use of the vee principle when the shape of the work will permit it. If four points are needed in the same plane for proper support or location of a rough surface, be sure that one of these points is movable to compensate for inequalities in the surface. If a finished surface is to be used for locating, the pads on which it rests should be as small as possible (consistent with sufficient surface for proper clamping) and should be easily accessible for cleaning.

2. Clearances around the casting: These should be made ample to accommodate variations in the work. As a rule one-quarter inch on all sides of the rough piece is sufficient, on

medium-sized work, such as that machined on horizontal turret lathes. On larger work for the vertical turret lathe or boring mill, one-half inch is none too much and even a little more than this is safer, for larger castings vary greatly in size and the author has seen a fixture (within the last two years) which had to be machined to obtain sufficient clearance, although it was designed to give one-half inch clearance all around. Needless to say this was made for a large casting, and was used on a boring mill.

3. Clamping: Methods of clamping are many but the plain strap is perhaps more used than any of the others, partly on account of its simplicity but also because it is very efficient. Hook-bolts are good if overhang is not too great, but they are worthless if not well backed up. They are also rather expensive. It is advisable to so design the clamping devices that they can be rapidly operated so that valuable time may not be lost in setting up or taking down the work. It is not considered good practice to use clamping screws which are tapped into a cast-iron body. It is much better to make these in the form of a stud with a nut and washer on the clamp. This brings the wear of the clamping action on steel instead of cast iron.

4. Chips: Provision should be made for the removal of chips so that accumulations of these will not cause trouble. Accessibility to bearing or locating surfaces is important so that cleaning can be readily accomplished. When fixtures are of the pot variety cored openings may be arranged for this purpose. When fixtures are designed for use on vertical boring mills this point is of great importance and must be carefully considered in the design.

5. Rigidity: On the horizontal type of machines this point is largely dependent on the overhang from the spindle. Therefore the fixture should be designed in such a way that this will be as short as possible. Ribbing may also be used when needed to give additional stiffness, especially around the clamps, as at these points the strains are excessive and changes may take place in the work itself unless provision is made to neutralize

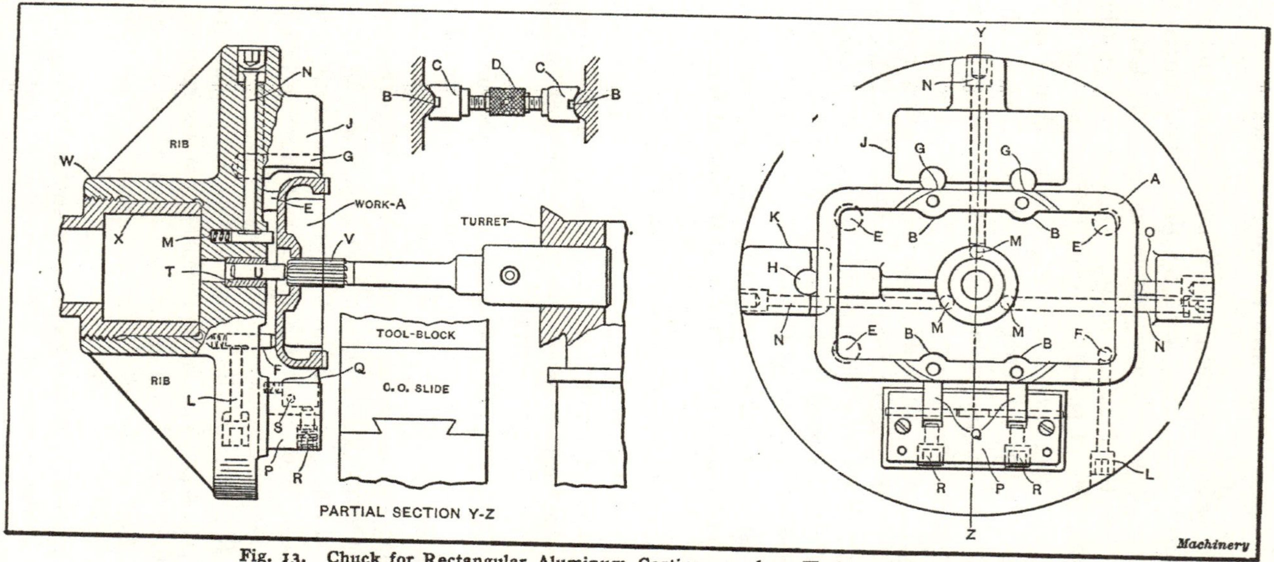

Fig. 13. Chuck for Rectangular Aluminum Castings used on Horizontal Turret Lathe

distortion. On the vertical type of machines fixtures should be made of generous proportions to resist the heavy cuts which these machines commonly take.

6. Safety: The operator should be considered at the time the designing is done and not afterward. Projecting lugs, set-screws, etc., should be avoided as far as practicable, especially when their location is at the outer portion of the fixture. Lugs can be easily made so that they have round corners which will not catch in the operator's clothing, and set-screws of the hollow variety may now be obtained in many different forms suited to almost all conditions.

7. Cost: This should be to a certain extent proportional to the accuracy required in the finished work and also to the number of pieces of one kind which are to be machined. A very elaborate and costly fixture should not be designed for a case calling for only a few pieces as this cost distributed on the work would greatly increase the cost of each piece. When a large number of castings are to be handled, however, the first cost may be almost overlooked and every improvement used which will tend to decrease the production time.

In the examples given herewith attention will be called to some of the important points in design and construction, faults will be emphasized, and suggestions made. It should be borne in mind that the examples illustrated have been selected with a view to simplicity rather than complexity of design, so that basic principles will be readily grasped.

Chucking Fixture for a Rectangular Aluminum Casting. — The work shown at *A* in Fig. 13 is an aluminum cap for an electric generator. The casting is somewhat rectangular with round corners and is of rather thin section with practically no reinforcement in the shape of ribs. No work has been done upon it previous to this setting so that the casting is in the rough. The machining necessary for this series of turret lathe operations consists of boring, reaming and facing the inner hub; facing the rim and sizing the segmental circular tongues on the face of the rim. An accurate job was required, and no distortion of the casting was permissible.

The work was placed in the fixture so that it rested on the three fixed pads *E* located at the corners, and a spring pin *F* acted as an adjustable support at the other corner. The coil spring under the pin is just strong enough to insure positive contact with the work without danger of springing. The set-screw *L* is of the hollow type and serves to lock the pin securely in position. The side and end locations of the work are determined by the pins *G* and *H* which are flatted off to form a knife-edge where they touch the casting. This arrangement causes them to sink in and prevents any tendency to pull out of the fixture. It will be noted that these pins are set into the blocks *J* and *K* which form a solid backing and prevent springing. A steel block *P* is screwed to a finished pad on the fixture body and is doweled in place. Two knife-edged clamps *Q* fit slots in the block and are pivoted on the pins *S*. They are forced into the casting by means of the hollow set-screws *R*, and have coil springs for the purpose of keeping them back when not in use.

In connection with these clamps, attention is called to the points on which they pivot; these are placed well back from the knife-edge so that the clamping action also has a pulling-in tendency. As a means of support for the center of the work, the three spring pins *M* are provided, these being arranged in the same manner as the pin *F*, and locked in position by the set-screws *N*. The body of the fixture *W* is well ribbed up at the rear and is bored out and threaded to fit the spindle *X*. A tool steel, hardened and ground bushing *T* is forced into the center of the body, and acts as a guide for the boring and reaming bars. The pilot *U* of the floating reamer *V* is shown in position.

After this fixture had been completed it was tested and was found unsatisfactory, due to the spring of the work at the clamping points *B*. In order to remedy this defect, a pair of special supporting jacks were made such as shown in Fig. 13. Two steel vee-blocks *C* were turned down and threaded with right- and left-hand threads, respectively, to fit the knurled collar *D*. These jacks were placed in position in the casting before it was placed in the fixture, and were tightened by the fingers, a piece

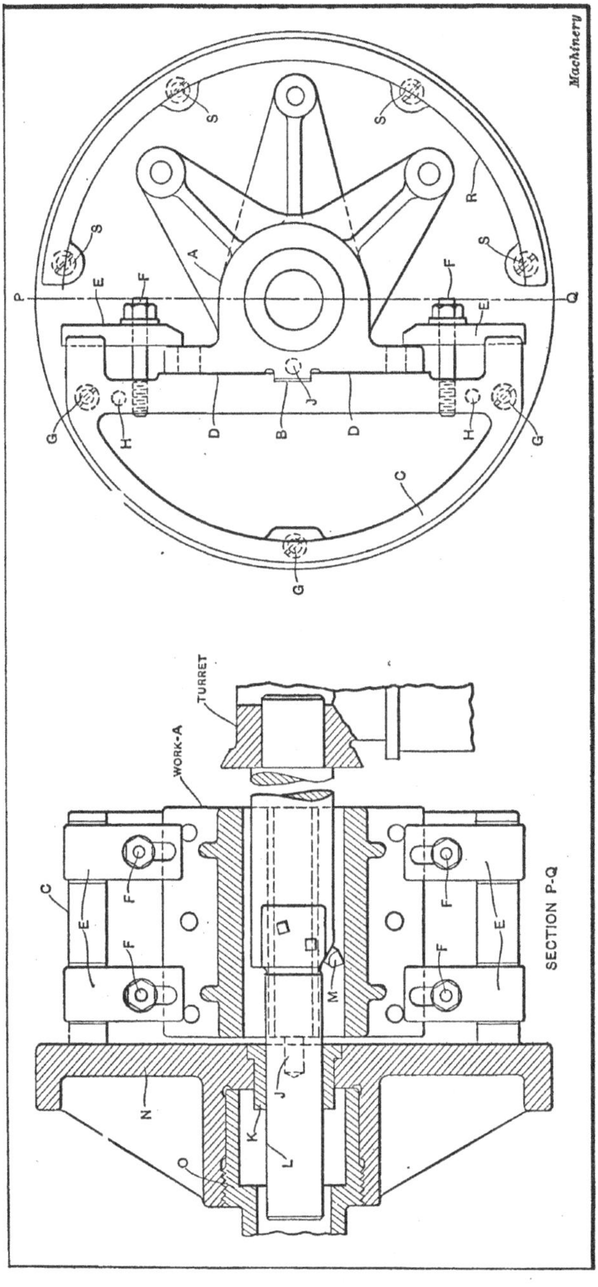

Fig. 14. Fixture for Boring Irregular Castings on Horizontal Turret Lathe

of drill rod being used for the final pressure. By the use of this device a perfectly rigid effect was produced and the work obtained thereafter was entirely satisfactory. In the design of this fixture as a whole, attention is called to the freedom from projecting set-screws, and other protuberances likely to catch in the operator's clothing.

Special Fixture for a Bearing Bracket. — The work *A* shown in Fig. 14 is a cast-iron bearing bracket of somewhat peculiar shape. This has been planed in a previous setting, on the base *D*, and the tongue *B* has also been machined to size. The body of the fixture *N*, made of cast iron, is screwed to the spindle *O* of a horizontal turret lathe. It was faced off in position on the machine so that the face would be perfectly square with the spindle. The semi-circular pot casting *C*, carefully located on it by the dowels *H*, is held in place by the screws *G* which enter it from the rear. The work *A* is located on the face of this casting by the tongue *B* which fits a slot cut to receive it. The clamps *E* are of steel and are slotted so that they can be pulled back out of the way when placing the work in position or removing it. The studs *F* are threaded to a tight fit in the casting *C*, and are provided with nuts and washers which bear on the clamps. In this construction the wear due to the operation of the clamps all comes on the steel of the screw rather than on the cast iron, this tending to make the life much longer, and the up-keep better. The lugs on which the ends of the clamps bear are slotted in order to prevent twisting while they are being tightened. These slots are not machined but are cast and afterward smoothed out with a coarse file. The pin *J* in the body of the casting simply acts as a longitudinal stop.

Partly as a protection to the operator and partly as a counterbalance, the segmental piece *R* is screwed to the body by the four screws *S*. This arrangement is valuable because the work is revolving at a fairly good speed (120 R. P. M.) so that good balance is important, and danger to the operator is avoided on account of the guard over the projecting lugs on the work. Attention is called to the accessibility for cleaning the locating surfaces and also to the ease with which a wrench can be used

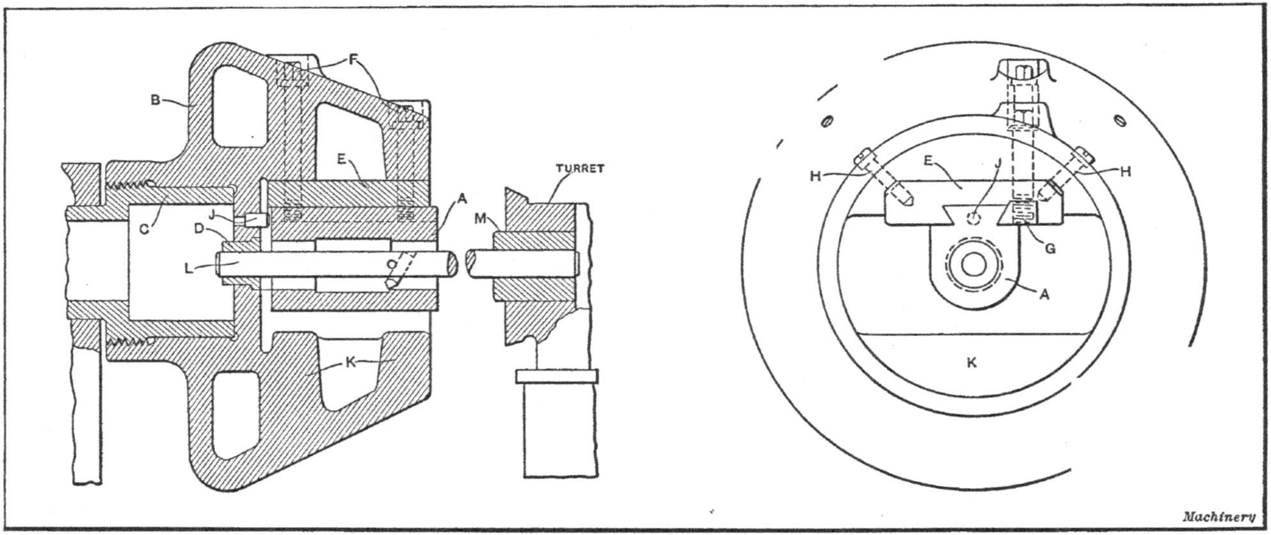

Fig. 15. Fixture for Boring Small Bronze Castings on Horizontal Turret Lathe

on the clamping nuts. A point in the design which could be improved is the amount of surface used for locating. Small pads under the clamping surfaces and a rim on each side of the tongue would have been ample and could have been kept clean more easily. A tool steel bushing *K* is forced into the body of the fixture and acts as a guide for the pilot *L* of the boring-bar, which greatly assists in preventing chatter while the tool *M* is cutting. The work accomplished in this fixture was satisfactory.

High-speed Fixture for a Bronze Bearing. — The bronze bearing shown at *A* in Fig. 15 has been previously machined on the dovetail portion, and it is of importance that the bearing should be in a fixed relation and parallel with this dovetail. As the hole is of small size and the work of bronze, it is necessary that the fixture should run at high speed in order to produce the work in a reasonable time. The body of the fixture *B* is made of cast iron and is fitted to the spindle *C* of a horizontal turret lathe. A steel locating block *E* is fitted to the dovetail of the work and is fastened into its seat in the body by the screws *H*. Two screws *F* are tapped into the dovetail gib *G*, so that they can be used to tighten the work in the fixture without chance of distortion. The pin *J* acts as a longitudinal stop. The lugs *K* are provided for the purpose of balancing the fixture so that it will run without vibration which might otherwise be caused by the high speed and spindle overhang.

It will be noted that the outside of the fixture is smooth, thus offering no danger to the operator. Obviously a socket wrench is used to tighten the gibs which secure the work in position. A tool steel hardened and ground bushing is forced into the body at *D* and acts as a guide to the boring-bar *L* which is necessarily small and needs support. The other end of the bar is held by the bushing *M* in the turret hole. This fixture is quite simple but the method used is a little out of the ordinary. It should be noted that the clamping device is efficient and does not tend in any way to distort the work. Its action is in two directions on account of the angle of the dovetail and therefore makes a positive location, as it crowds the work into

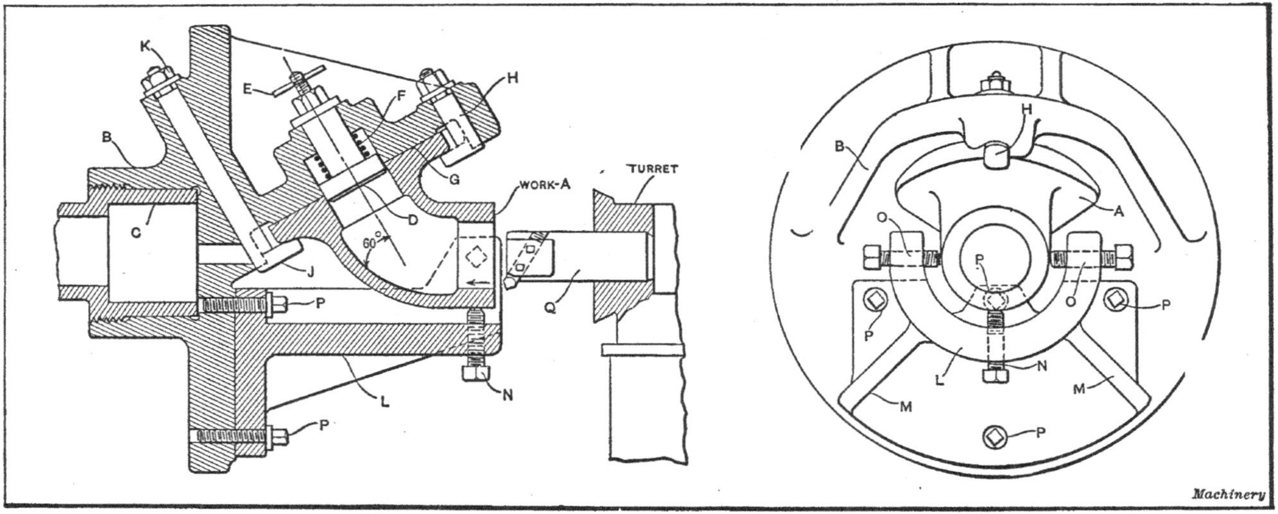

Fig. 16. Fixture for Machining Angular Cast-iron Fitting

the dovetail and draws it back at the same time. The fixture was used with satisfaction.

Fixture for an Angular Cast-iron Fitting. — The rather awkward piece of work shown at *A* in Fig. 16 is an angular cast-iron fitting which has been previously machined at *G* and had the hole at this end bored. The angle required between the two openings is sixty degrees. The machine to which this fixture is fitted is a horizontal turret lathe. The body of the fixture *B* is of cast iron and is screwed to the spindle *C* in the regular way. The angular pad which receives the finished portion *G* of the casting is machined to give the correct relation between the holes in the joint. A spring plunger *D* is located centrally in this pad, and serves to locate the work in relation to the previously bored hole. The pin *E* is used to pull the plunger back when placing the work in position. The hook-bolts *H* and *J* grip the piece by the flange and hold it firmly against the locating pad. It will be noted that *J* passes through and is operated by the nut and washer at *K* on the back of the fixture. In order to insure rigidity to the forward portion of the piece where the work is being done, a segmental casting *L* is used, which is secured to the body by the four screws *P*. The set-screws *N* and *O* are cup-pointed, and greatly assist in keeping the work rigid at this end. It will be seen that the bracket is well ribbed up at the points *M* to secure additional stiffness. A plain boring-bar *Q* is held in the turret.

Simple Arrangement Using Four Jaws. — The motor bracket casting shown in Fig. 17 is of cast steel and has been previously machined on one of its faces. Four jaws in a chuck of the independent type are used in this setting of the work, and the machine employed is a vertical turret lathe. The bracket which is to be machined is located by the two jaws *B* and *C* which are left set to fit the work. The function of these jaws is to take the place of a vee-block and the other jaws are used for the purpose of holding the work securely. The shape of this piece of work being rectangular, it is feasible to use this method for locating, by allowing them to act as a vee. In action, *E* and *D* are brought up alternately until the work is securely held. The surface of the work which has been previously machined is supported by the steel buttons F which are positively located in the jaws. This method is adaptable to work which comes along in small lots but it is open to objections. The principal one of these is the possibility of the operators shifting one of the vee-jaws unconsciously, thereby ruining valuable castings. Another is the possibility of variations in the squareness of the machined face with the gripping surface, which will naturally result in work which is not absolutely true. But as a makeshift method when only a few pieces are to be machined, it is satisfactory in the majority of cases unless a very careless workman is employed.

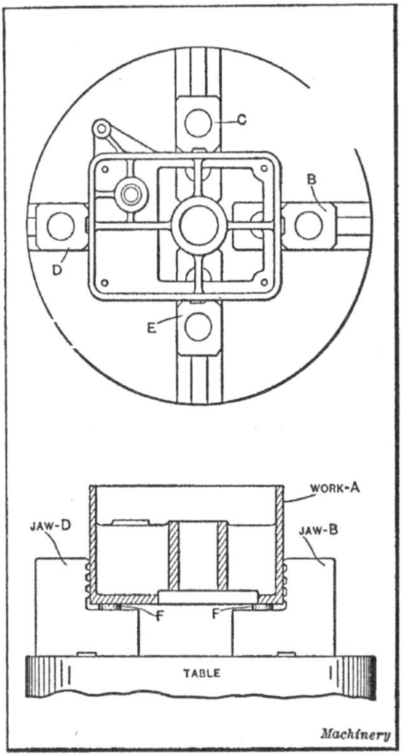

Fig. 17. Use of Four-jaw Chuck for Holding Rectangular Work on a Vertical Lathe or Boring Mill

Pot Fixtures. — The steel casting *A* in Fig. 18 is a piece of work which is handled in the rough, no previous machining having been done on it. This casting is of large size and being of steel is subject to variations in size and shape. As this is the case it is necessary to make suitable provision for these so that compensation may be obtained to suit the various conditions. The vee-principle is made use of in the locating device as far as the cylindrical portion is concerned, the set-screws *K* being adjusted to suit the casting, so that it is centered from the rough exterior. The pot casting *B* is of cast iron and is centered by the plug *P* in the center of the table. This plug is forced into the pot casting at *O*. The work is dropped into the fixture and forced over against the vee-screws *K*, by the square-head set-screws *J* which are tapped into the lugs *H*, and come against the open portion of the work at *F* and *G*. Support is obtained by the screws at the three points *D*, two of which are movable and the third fixed. The movable points are as shown in the lower view, and are adjusted by means of a piece of drill rod, so that the entire casting can be tipped one way or the other to compensate for inequalities in the rough casting.

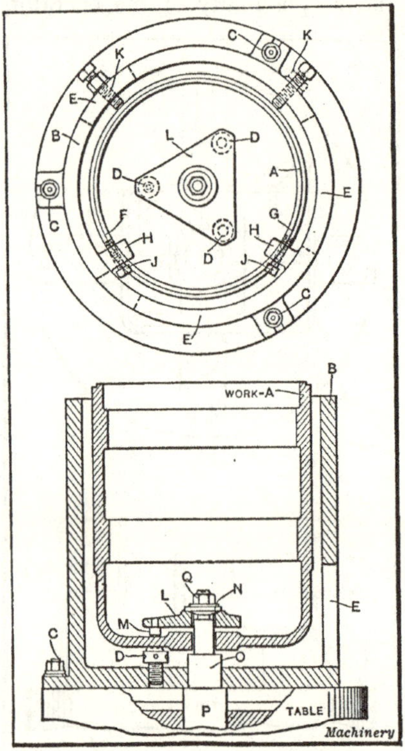

Fig. 18. Pot Casting Fixture for Holding Steel Castings on Vertical Turret Lathe

The steel triangular plate washer *L* is fitted with three clamping points *M* which bear on the inside of the rough casting and are equalized so that they all get the same amount of pressure, by the spherical washer *N*, operated by the nut *Q*. The entire fixture is held down on the table of the machine by the T-bolts *C* which fit the table T-slots. This fixture gave results which were satisfactory, but the setting-up time required was rather long. The openings *E* shown in the illustration permitted access to the jack-screws and also allowed cleaning to be easily accomplished. The projecting set-screws are a rather bad feature, as they are dangerous to the operator, but as the speed was not excessive, no trouble was experienced with them. This defect could have been very easily remedied.

The casting *A* shown in Fig. 19 is of steel and is handled in the rough, no previous machining having been done. This work could have been handled in a three-jaw chuck combination, had it not been for the projecting lugs on the casting, and even these could have been taken care of by cutting away the sides of the jaws. A pot casting method was decided upon instead of this, however, on account of the greater rigidity. The body *D* is of cast iron and is fastened to the table with the four T-bolts *E*. On one side of this body are two pads *B* which are planed out to form a pair of vee-blocks in which the cylindrical portion of the casting locates. It is forced into position by the screws *F* in the wall of the pot casting. The entire fixture is located on the table by means of the steel plug *H* which fits the table center hole and is forced into the fixture at *J*. A driver is provided in the screw *G* which bears against the hub of the casting and takes the thrust of the cut. Although this is a very simple fixture it illustrates the principle of the vee-block to advantage and should therefore be carefully noted.

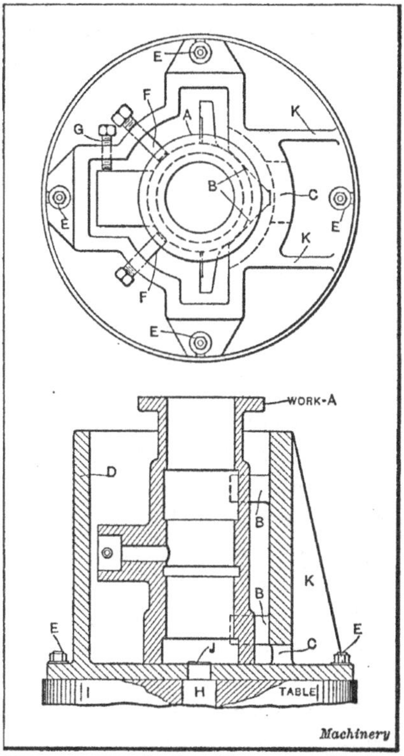

Fig. 19. Pot Casting Fixture for Holding Irregular Steel Castings on Vertical Turret Lathe

CHAPTER VIII

TAPER BORING AND TURNING ATTACHMENTS

Machines used for Taper Turning and Boring. — The proposition of accurately machining male and female tapered surfaces is one of almost daily occurrence in every factory, while the tapers required are of every degree of inclination. The materials on which the work is to be done are also varied, both in their general form and in their composition, ranging from steel or brass bar stock of small diameter to cast iron or steel castings of great size. Conditions governing the work are widely different, as the number of pieces needed obviously makes a difference in the method of handling. When only one or two are required, and the size of the work is not prohibitive, the engine lathe is most frequently used, several well-known methods of generating the taper being possible on this machine, *viz.*, setting over the tailstock to the correct angle, when the work is of such a nature that it may be held on centers; using the compound rest with hand feed; and using the taper attachment with which nearly all modern lathes are equipped and which is too well known to need description.

There are also occasional instances where the lathe may be used for manufacturing work of this kind in large quantities, by means of special attachments, although this is usually applicable to conditions requiring no other machining operations except the taper. As a general thing when the number of pieces is sufficiently large to warrant it, the work is performed on the horizontal screw machine or turret lathe, the vertical turret lathe or the vertical boring mill. Many ingenious schemes for generating tapers on these machines have been devised, the construction of a number of which will be described and illustrated in this chapter.

Taper Turning Devices for Bar Stock. — On turret lathes or screw machines equipped for bar work, there are various devices for turning a taper on the bar. These tools are in many instances patented, and may be purchased from the manufacturers. Obviously there are such a number of these that it is out of the question to attempt to describe each one. Detailed information may be easily obtained, on request, from the manufacturers.

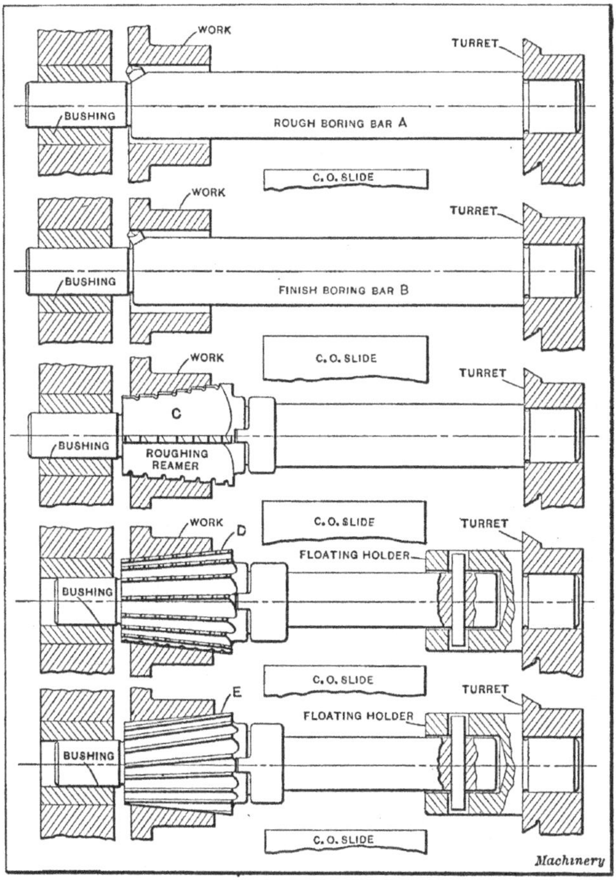

Fig. 1. Typical Boring Tools and Reamers for Taper Holes

Method of Finishing a Taper Hole without Generating the Taper. — Before taking up the subject of generating devices for taper work, let us first consider a method much used in turret lathe practice and one which may be depended upon to give very satisfactory results, when absolute accuracy is not essential. When the tools are properly taken care of, good commercial work may be turned out by means of the tooling shown in Fig. 1. It will be noted that all the tools used are piloted in a bushing located in the chuck. The first tool used, shown at *A*, is a plain boring bar which serves to rough-bore the hole, thereby producing an approximately true generated straight hole. The second tool *B* is a finish-boring bar which brings the hole to about the required size for the small end of the taper. The next tool *C* is a roughing taper step reamer which removes the larger part of the stock left in the hole, and leaves the work in the form of a series of grooves or steps with the angle of the correct inclination. A roughing taper reamer *D* is next used in a holder so made that the rear end of the reamer will float. It will be noted that this reamer is straight-fluted, but that a left-hand spiral groove with about $\frac{3}{4}$ inch lead is cut the entire length of the tool. This serves to break the chip and makes for much easier cutting, also having a tendency to prevent pulling in.

The hole is sized with another reamer of the floating type *E* which may be either straight-fluted or made with a left-hand spiral of five to seven degrees, depending on the angle of the taper. The method shown here will not give as accurate results as may be obtained by generating the taper, but the sizing of the hole may be kept very nearly correct with little trouble, although slight variations in concentricity are bound to occur. One of the greatest objections to this manner of handling taper work is that the operator does not keep his tools up properly, and by being careless in regard to this matter, he leaves the reamers to do the most of the work and the results are therefore disastrous on account of the unequal wear on the reamer.

Taper Attachment for Producing a Conical Surface on the Engine Lathe. — Fig. 2 shows an attachment fitted to the engine lathe for the purpose of producing the proper angle *C* on the head casting *A*. In this case the work is held in special jaws *B* which grip the interior of the casting as shown in the upper part of the illustration. The cross-slide is equipped with a special tool-block *D* in which the tool *E* is held. The casting *H* was planed on its under side to fit the inner ways of the lathe and was clamped in position by means of straps not shown. Two steel plates *K* and *L* act as guides by which the proper taper is formed. These plates are hardened and the edges of the slot were ground parallel after assembling, to insure accuracy. A bracket F, fastened to the cross-slide, carries a pivoted steel block *G* which travels in the slot, thereby controlling the movement of the carriage, and pro-

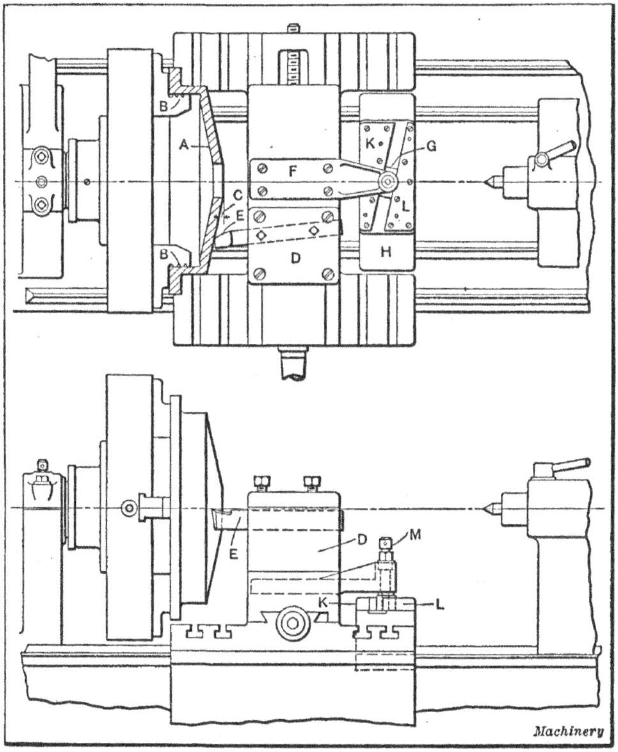

Fig. 2. Engine Lathe Attachment for Turning Bevels

ducing the desired taper. An oiler *M* acts as a gentle reminder that surfaces subject to friction are in occasional need of lubrication. The inner surface of casting *A* was machined on the same lathe in another setting, another set of forming plates being applied to the casting *H* to produce the required taper. This method of handling gave very satisfactory results.

Taper Attachment for a Small Hand Screw Machine. — A small brass cock, shown at *A* in Fig. 3, is a good example of an outside generated taper. The stem of the cock *B* is held in the special jaws *C* of a two-jawed chuck *D*, this being obviously screwed to the spindle of a small hand screw machine. The taper turning attachment is entirely self-contained, and indexes with

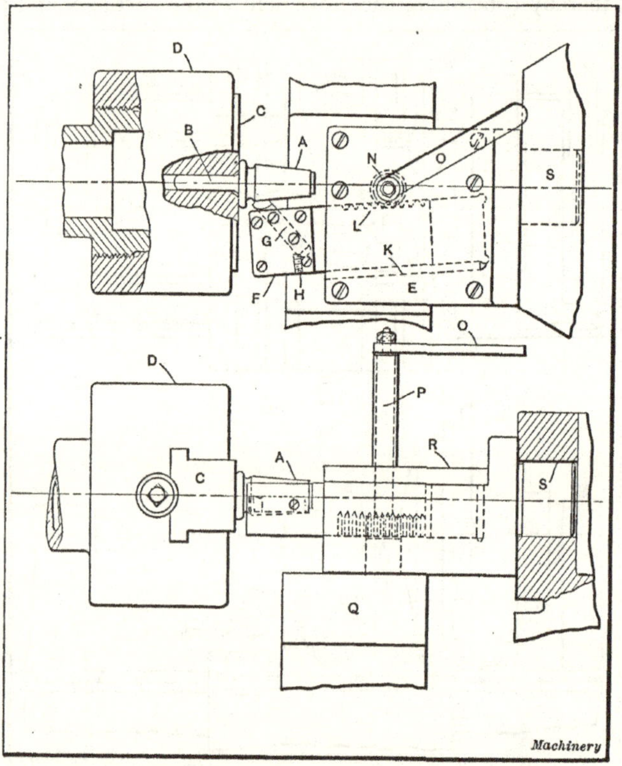

Fig. 3. Taper Attachment for Small Hand Screw Machines

the turret. The entire attachment is made of steel with a shank *S* which fits the turret hole. The body *E* is carefully fitted on its under side to obtain a bearing on the steel block *Q* which is fastened to the cross-slide. This support is of considerable help in taking up vibration and thereby preventing chatter. The slide F fits a slot in the fixture which has been planed to the proper taper and the gib *K* acts as a take-up for wear. The cutting tool *G* is of rectangular section and accurately fits a slot in the front end of the taper slide. The headless set-screw *H* assists in setting the cut to obtain the proper diameter. A rack *L* is cut along one side of the slide and meshes with the pinion *N*, the shank *P* of which runs up through the body of the fixture and is operated by the lever *O*. A cover plate *R* is carefully fitted and keeps the parts in position. Tools of this type are much used on small brass work and the work accomplished by them is excellent where very little stock is to be removed. They are built to generate a certain specified taper and can be used for no other.

Attachment for Generating Small Taper Hole in a Motor Cycle Flywheel. — The rather complicated little attachment shown in Fig. 4 was built for a final finishing cut in the taper hole *D* of the motor cycle flywheel *A*. In spite of the fact that the attachment itself is inclined toward multiplicity of moving parts, its action was so satisfactory that a duplicate order was received a few months after the original tool had been built. It will be noted that the jaws *B* of the three-jawed chuck *C* grip the work on the inside of the flange, and hold it far enough away from the chuck to permit back cutting on the hub and flange, thereby permitting the work to be finished in one setting. The body of the attachment *E* is made from a piece of round steel stock beveled on the front end and with the shank F turned at its rear end to the proper diameter to fit the turret hole. The taper slide *G* fits an angular slot cut in the body of the attachment and is reamed at its front end to receive the shank of the cutting tool *N*. This tool is forged to the shape shown and is carefully ground to gage. As the amount of metal which this tool removes is very slight, it requires regrinding only at long intervals. A headless set-screw *T* secures it in position.

The bell-crank *K* is pivoted at *M* and the hardened steel rollers *L* and *U* are located at the two ends. The roller at the forward end operates the slide by its action in the slot *H*, while the roller at the other end enters another slot *O* in the operating pin *P*. A teat-screw *V* enters the spline *W* cut in the under side of the operating pin, thereby preventing it from turning. The small knurled handwheel *R* contains a little finger handle *S* which is used to revolve the screw *Q*. The rod *P* is tapped out to receive this screw, and obviously is moved forward or backward by its action, the motion being carried forward through the bell-crank to the operating slide.

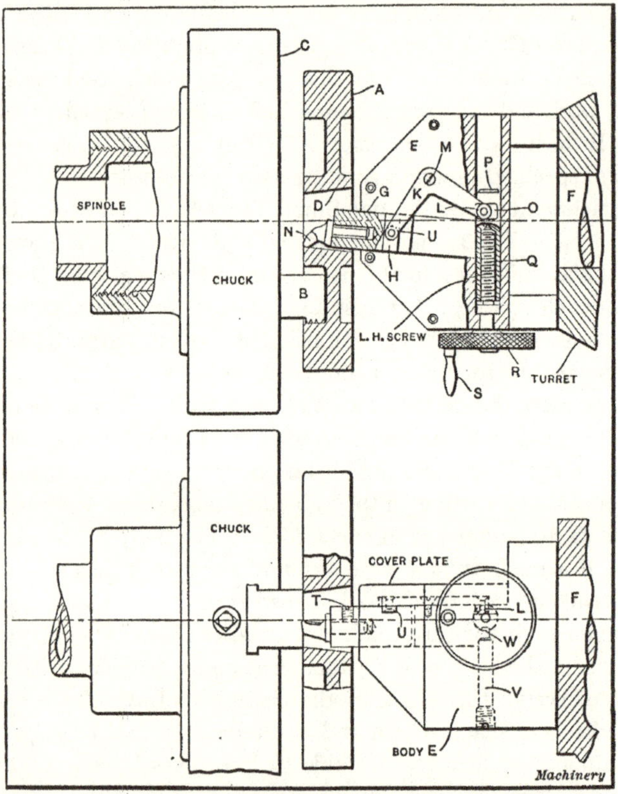

Fig. 4. Attachment for Generating Small Taper Hole in a Motor Cycle Flywheel

Turret Lathe Taper Attachment for the End of a Piston. — The automobile piston shown at *A* in Fig. 5 has been finished on the outside but the end has not been formed to the required taper. It is held in a special spring chuck *B* which is closed in on the end by the tapered screw collar *C*. In this instance there were several conical-headed pistons to be taken care of, the angle of the cone varying slightly in each case. The turret lathe selected for use in this operation was of a standard make, and the longitudinal movement of the cut-off slide was controlled by the screw *R* engaged with the nut *V* on the under side of the slide. This screw was operated by a handwheel and was not coupled up with

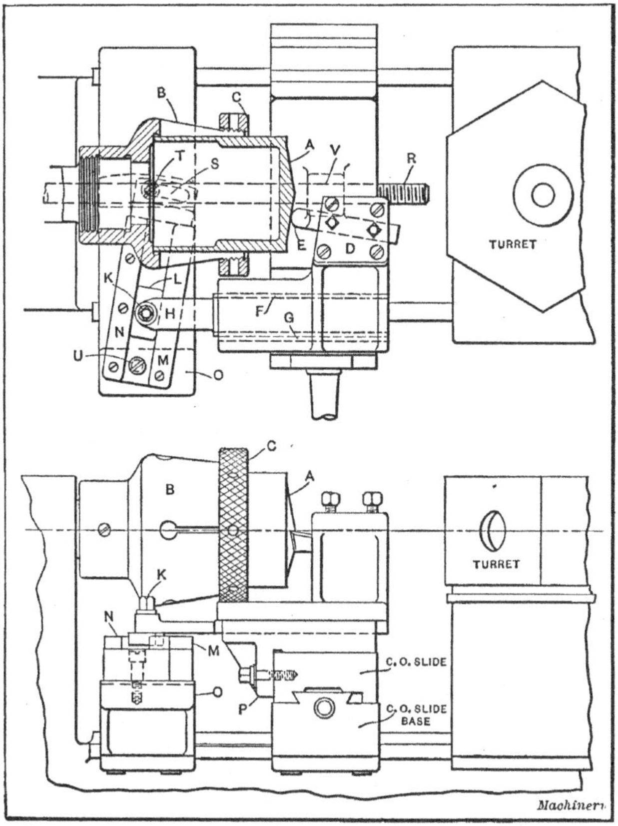

Fig. 5. Turret Lathe Taper Attachment for Machining the End of an Automobile Engine Piston

the feed mechanism. It was used principally to move the slide back and forth along the ways to any desired location.

It will be seen that in this instance any sort of floating action in a longitudinal direction was out of the question and it was therefore necessary to design a special tool-block *D* having a dovetail slide *F* and an extension *H*, at the end of which the hardened and ground steel block *L* was located, pivoting on the screw pin *K*. A swivel block containing two parallel plates *N* and *M* may be swung on the shouldered screw *U*, to suit the various angles. The curved slot *S* permits the necessary movement, while the binder *T* secures it firmly. The swivel is mounted on the bracket *O*, which is gibbed to the ways in such a way that it may be moved to any desired location. When the attachment is used the cut-off slide cross feed is thrown into gear and the angularity of the swivel block determines the movement of the tool *E*. As a point in design, attention is called to the way in which the tool-block is carried over the edge of the slide at *P*, for the purpose of obtaining rigidity and preventing any chance of side slip. The author knows of a number of instances where attachments of this kind have been used with very gratifying results.

Swivel Cut-off Slide Attachment. — The bevel gear shown at *B* in Fig. 6 is held in the three-jawed chuck at *C*, by means of soft jaws, and the tools *M* and *N* are used for roughing and finishing the face of the gear, the spacing of the tools being such that the finishing tool takes up the work as soon as the roughing tool has completed its cut. This entire mechanism is special.

The carriage *S* is gibbed to the ways in the usual manner and upon this carriage is mounted the swivel slide arrangement *D*. This slide may be swung to any angle within its range and securely fastened. It will be noted that the feed-screw *P* meshes with the wormwheel *Q*, and the movement is transferred through the spur gears *T* and *R* to the shaft *K*. At the inner end of this shaft the pinion *G* meshes with the bevel gear *H* on the upper end of which is the spur gear F. This spur gear engages the rack *E* cut along the inner side of the slide, thus giving the necessary feed movement. The tool-block *L* is held in place by screws which pass down through it into steel shoes in the T-slots *O*. The operation of this mechanism is so apparent that no comment is necessary.

Taper Attachment for a Bevel Pinion. — The bevel pinion *X* shown in the lower portion of the illustration, Fig. 7, has been bored with a taper hole and the back side faced in a previous operation. A keyway *Y* has also been cut for driving purposes. The equipment shown was designed for a large factory manufacturing bevel gears and pinions and the taper turning device shown is so arranged that it may be used for a variety of angles.

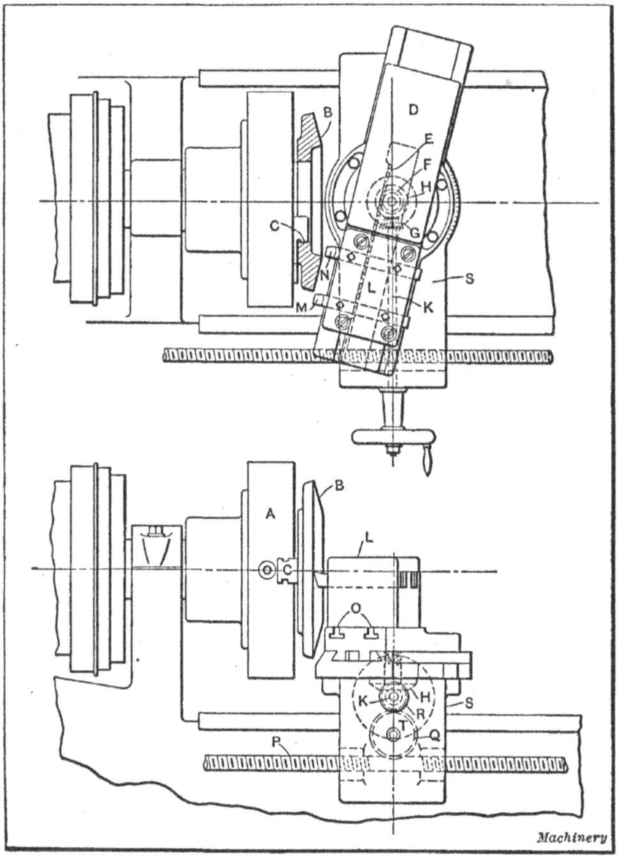

Fig. 6. Adjustable Cut-off Slide Attachment

A number of taper bars, such as that shown in detail in the upper part of the illustration, were made to suit the different conditions.

The spindle nose-piece *W* contains a tool-steel arbor pilot supported at *V* in the fixture bushing. The nut *Z* is simply used to release the work after the machining operation has been performed. A cast-iron adapter *L* is screwed to the turret face, and on this is mounted the body of the fixture *M*. The cutting tool *B* is held in the sliding tool block *A*, which is scraped to a nice sliding fit, and has a taper gib provided for adjust-

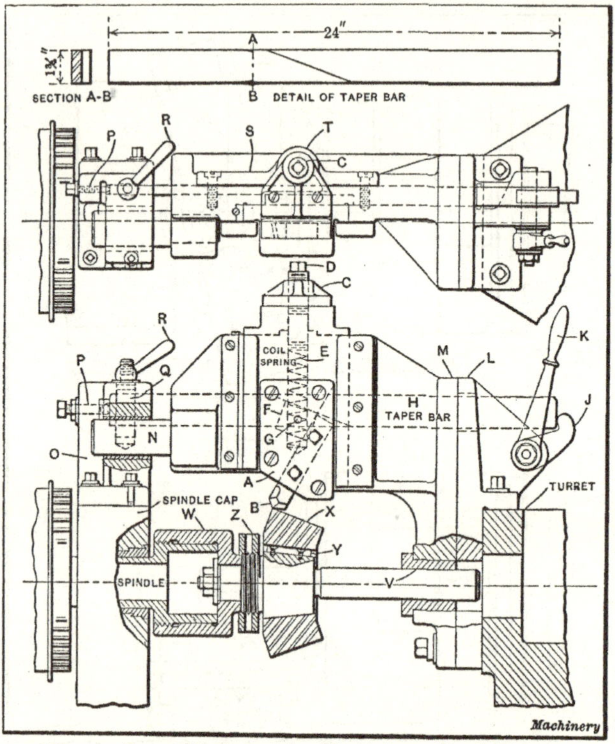

Fig. 7. Taper Attachment for Machining a Bevel Pinion

ment. A tool-steel block *F* is pivoted to the back of the slide on the pin *G*, allowing it to adapt itself to the angle cut on the taper bar.

Referring to the upper view it may be seen that the plate *S* forms a cover for the open side of the fixture, and that it contains the long boss *T* which holds the spring *E*. This spring thrusts against the end of the screw *D*. The bracket *C* is fastened to the top of the slide and is tapped out to allow adjustment to the spring by means of the screw. A bracket *O* is fastened to the spindle cap and contains a bronze bushing which acts as a guide for the pilot *N*. The stop screw *P* is used for longitudinal adjustment of the taper bar *H*. The lever *K* is used to force the taper bar forward by means of the

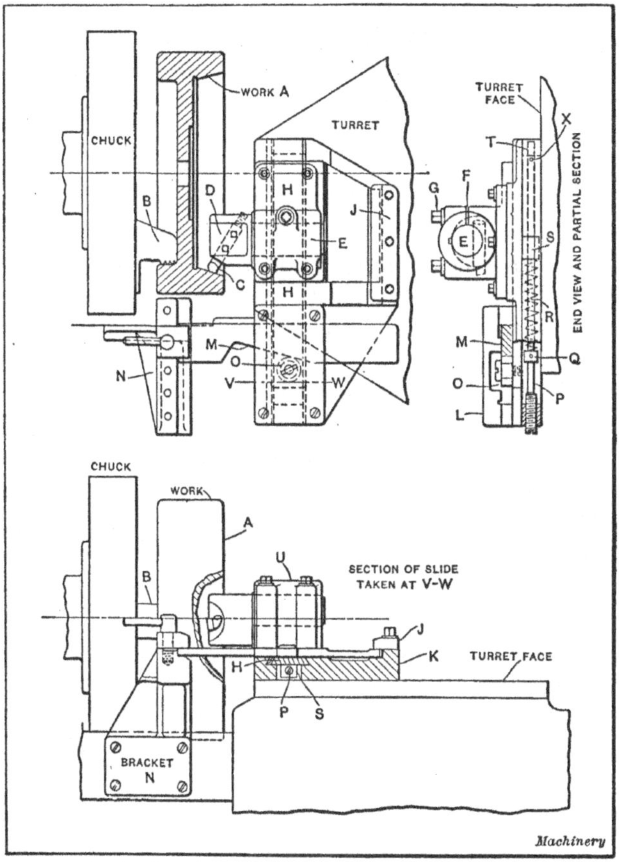

Fig. 8. Exterior and Interior Taper Turning Device

rocker *J*. The stud *Q* is slotted to receive the forward end of the taper bar, and when this has been brought forward by the lever until it strikes the end of the screw *P* the binder lever *R* prevents any backward movement of the bar. The adaptability of this attachment for various tapers is one of the good points of its design and the results obtained by its use are rapid and thoroughly satisfactory.

Exterior and Interior Taper Turning Device. — The device shown in Fig. 8 is adapted for use on a turret lathe having a flat surface instead of the usual box-shaped construction. A taper bar is used in this instance also, which is cut away at *M* to the desired taper. The attachment may be arranged for either inside or outside tapers, but it is shown in this instance at work on the clutch taper of the piece *A*, this being held by the inside in the chuck jaws *B*. The bed or body of the fixture *K* is fastened to the face of the turret and is dovetailed to receive the slide *H*. A lug *S* on the under side of this slide receives the thrust of the spring *R* (shown in the end view section). A screw *P* forms an adjustment for the compression of this spring through the collar *Q* which is made removable and can be transferred to the other end of the rod at *X*, when it is desired to use the attachment for outside tapers. The roll *O* is fastened to the slide and in its contact with the taper bar produces the required taper. The tool-holder *U* is split along the side at F and is bored at *E* to receive the shank of the tool bar *D*. The binder screw *G* is used for clamping. A cast-iron cover plate *L* is fitted so that the pads shown on its under side allow free movement to the passage of the bar. The bracket *N* is fastened to a pad on the side of the bed and is cut away at the top to the proper height so that the taper bar *M* will rest upon it. It is clamped in position by the binder shown, in order to prevent any chance for retrograde action. This attachment has been very successful and is adapted to a wide range of casting work.

Attachments for Vertical Turret Lathe and Vertical Boring Mill. — Fig. 9 shows the simplest of conditions which are met with in vertical turret lathe practice, and the method of handling requires no special attachments, the swivel slide of the main head being sufficient to take care of the taper boring, the hole being finally reamed to size by a floating taper reamer. The work *A* is a cast-iron hub and it is held in the special jaws *B*.

The work is centered by the steel inserted jaws *D* and the set-screws *C* are simply used to prevent vibration. The roughing bar *E* is first used to generate the taper and it is followed by the finishing bar F. Then the rough- and finish-facing tools *G*

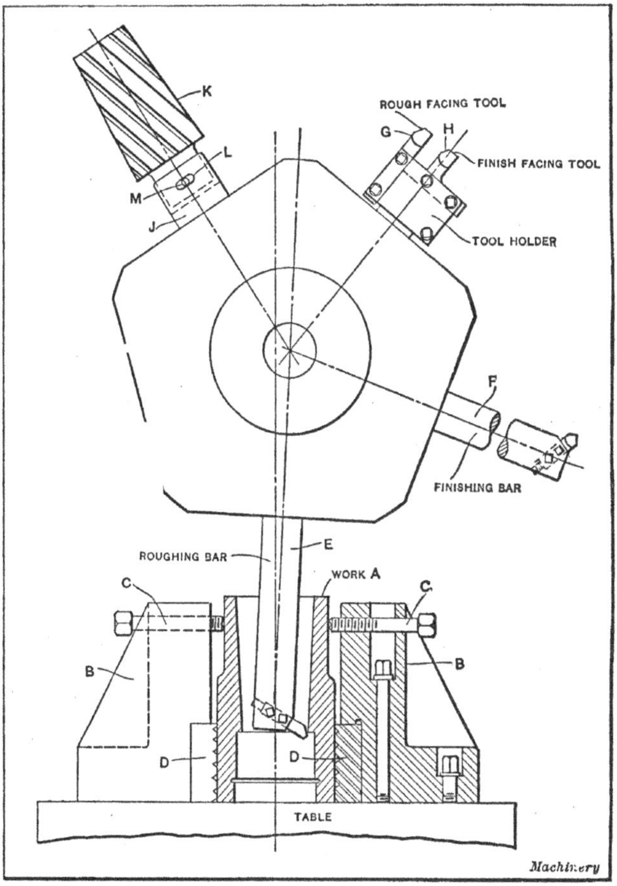

Fig. 9. Attachments for Vertical Turret Lathe and Vertical Boring Mill

and *H* face the work, after which the floating taper reamer *K* is used to size the hole. It will be noted that the upper end of the reamer is flatted and enters a slot in the holder *J*, the pin *M* acting as a driver and the slot *L* allowing lateral movement.

Special Gearing used to produce Tapers. — The arrangement shown in Fig. 10 is not adapted to all conditions but may be used when the required angle is not too acute to permit the use of the proper gear ratio. A piece of work such as that shown at *A* may be handled to advantage by this method.

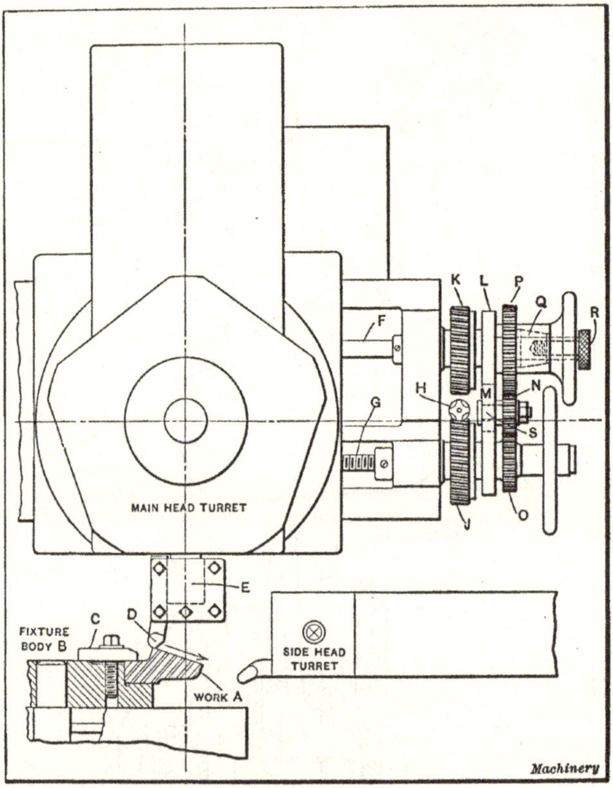

Fig. 10. Illustrating Use of Vertical and Horizontal Feed Combination to produce Tapers

The strap *L* is slotted at *M* to receive the stud *S*, which acts as a support for the idle spur gear *N*. The lower spur gear *O* is keyed to the shaft, while the upper gear *P* is thrown into use by the clutch mechanism *Q*, by the action of the knurled screw *R*. Obviously the gear ratios between *P* and *O* must be so proportioned that the combination of the horizontal and vertical feed movements will produce the required angle. Attention is called to the fact that the power feed worm *H* is thrown into mesh with the gear *J* on the horizontal feed-screw *G* when the attachment is to be used, but it will be seen that the operation of the feed works is not disturbed by the arrangement shown, the gear *P* running idle unless the clutch is thrown in.

The work *A* in the instance shown is held by straps *C* on

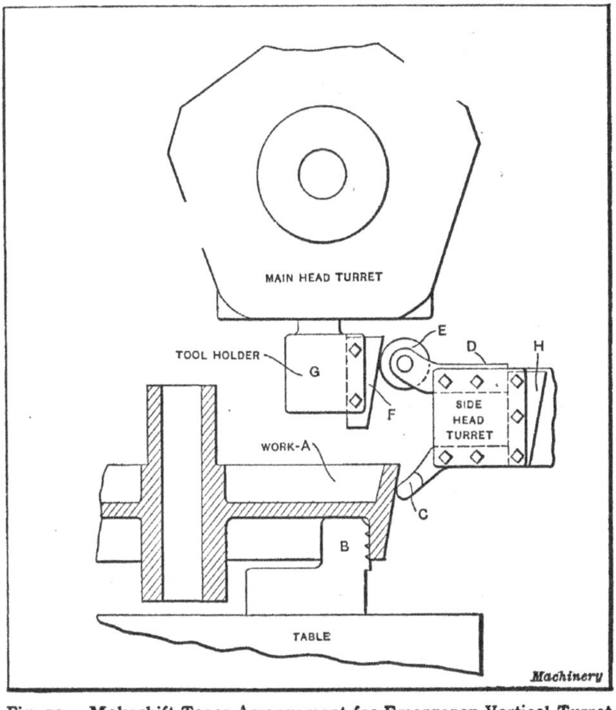

Fig. 11. Makeshift Taper Arrangement for Emergency Vertical Turret Lathe Work

the special fixture body *B*. The tool *D* is held in the toolholder *E* and follows the angle generated by the gearing. When it is required to produce an angle such that spur gearing cannot be obtained to give the exact taper, the nearest gears obtainable may be used, and the swivel slide of the main head can be set over to compensate for the variation in the gearing.

Makeshift Taper Arrangement for Vertical Turret Lathe.— Fig. 11 shows a method of setting up a vertical turret lathe for a rush job, consisting of a few cast-iron male clutch members shown at *A*. The work is held by the jaws *B* on the inside of the rim and the tool *C* is held in the side-head turret. A steel plate *F* cut to the required taper is held in the toolpost *G* in the main-head turret. A roll holder *D* is fastened in the

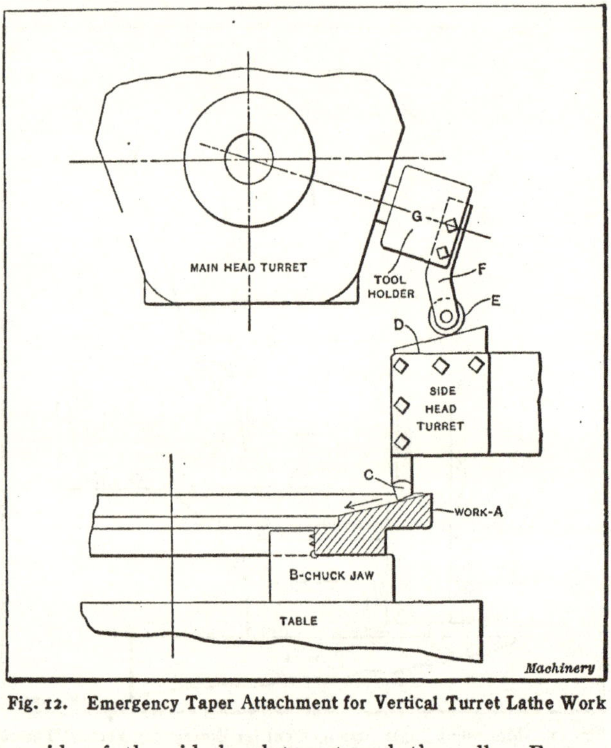

Fig. 12. Emergency Taper Attachment for Vertical Turret Lathe Work

upper side of the side-head turret and the roller *E* comes in contact with the tapered plate and thereby controls the movement of the tool. A flat angular sweep tool *H* is used for finishing the work. In using this arrangement it is only necessary to lock the main-head turret in the proper position and bring the roll *E* against the forming plate. After this the down feed of the side-head is thrown in and the roll crowded against the plate by the transverse feed crank on the apron. This method is very good for a short job and the machine may be quickly set up. Several roll holders of this kind will be found useful adjuncts to the tool equipment of the vertical turret lathe.

Another instance of a short rush job is shown in Fig. 12, the work *A* in this instance being held by the inside in the chuck jaws *B*. The tool *C* is held in the side-head turret and is forced down the angle by the contact of the angular plate *D* with the roll *E*. The shank F which holds this roll is secured in the tool-holder *G*, in one of the side holes in the main-head turret. When this arrangement is used the transverse feed of the side-head is thrown in and the plate *D* crowded against the roll by means of the vertical feed crank on the side-head apron. It will be readily understood that this arrangement and that shown in Fig. 11 are not to be considered in the light of attachments for taper turning, but they are given as instances of methods which may be used for short jobs, where no taper attachment is available. It is evident that these methods tie up the main head and prevent its use for cutting purposes while the taper is being formed. As this naturally increases the cutting time necessary to produce the work, the use of such an arrangement is advised only in cases where a few pieces are to be machined.

Angular Taper Attachment for Crowning Pulleys. — Fig. 13 is an arrangement which is used where a double angle is required, such as the crowned portion of the pulley *A*. In this case a set of special jaws *B* grip the work on the inside bead in the V-shape part of the jaw. The movement of the tool *C* is controlled by the forming plate *E*, which is cut to produce the angular movement required. This plate is fastened at each end to the bars *F* and *L*, and these bars are, in turn, secured in the upper and lower brackets *K* and *M*. The upper boss *G* is split and the binding screw *H* pinches the bar and holds it in the desired position vertically. The arrangement of the lower bracket is on the same principle. Both the upper and lower brackets are fastened to pads on the bed of the machine. When this attachment is used the T-slot *D* is cut along the entire length of the side-head slide so that the T-stud *O* which carries the roller *N* may be adjusted so that various diameters can be machined. This attachment is very satisfactory for work of this nature.

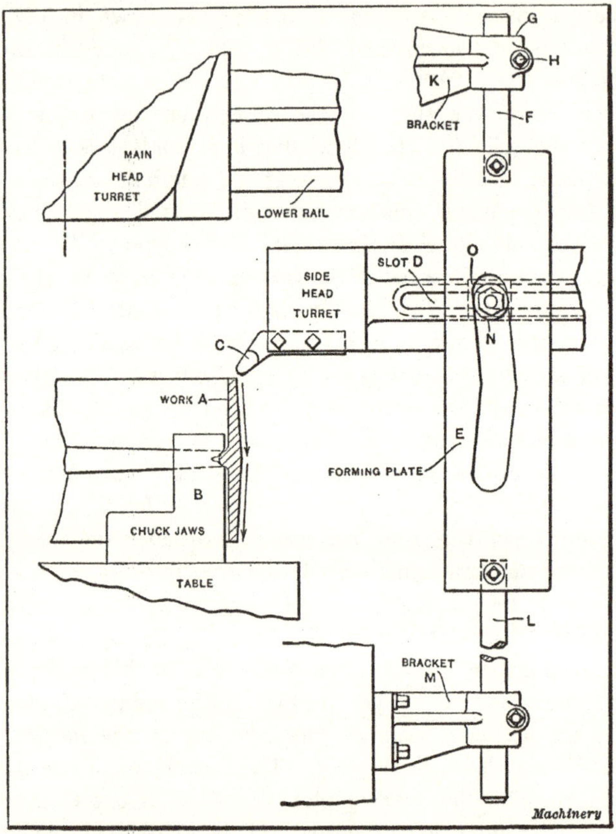

Fig. 13. Taper Attachment for Crowning Pulleys

Swivel Side-head Forming Attachment for the Vertical Turret Lathe. — Fig. 14 represents an attachment made by the Bullard Machine Tool Co., Bridgeport, Conn., for the Bullard turret lathe. The work shown in this instance at *A* is a large bevel gear which is held by the previously bored interior surface in the soft jaws *B*. The tool *C*, in its angular movement, is controlled by the inclination of the slot *D* in the circular swivel plate. This plate is graduated in degrees around its upper edge so that any angle may be easily obtained. The clamps *F* and *G* secure it in position after the setting has been made. The disk containing the slot is mounted on the plate *H* which is

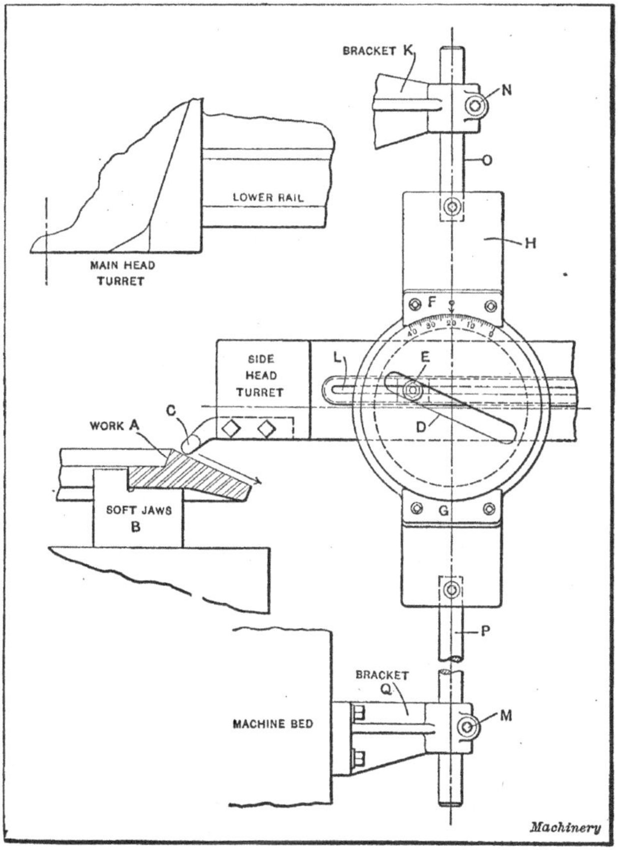

Fig. 14. Vertical Side-head Forming Attachment for the Bullard Vertical Turret Lathe

of circular section at the center to allow free access to the roll and block *E*. As in the previous instance a T-slot *L* is cut along the entire length of the side-head slide, thereby permitting various diameters to be machined. The bars *O* and *P* are secured in the brackets *K* and *Q* by the binders *N* and *M*, these brackets being secured to the bed of the machine. This attachment is adapted and may be used for many varieties of work and the results obtained are uniformly satisfactory.

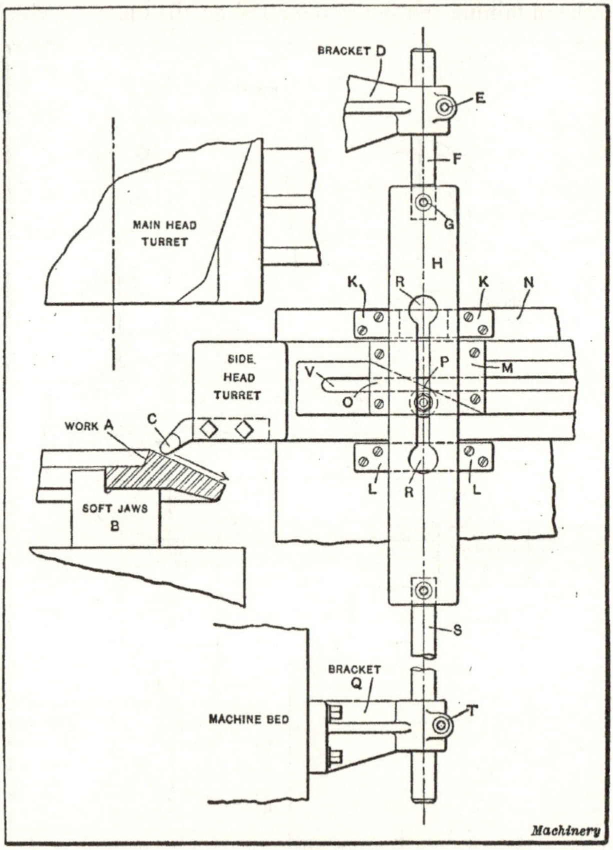

Fig. 15. Taper Turning Attachment for the Bullard Vertical Turret Lathe

Another Taper Turning Device for the Vertical Turret Lathe. — The device shown in Fig. 15 is also made by the Bullard Machine Tool Co., and is adapted to both angular and formed work, and therefore is more comprehensive in its uses than that shown in Fig. 14. The piece *A*, held in the soft jaws *B*, is the same as that previously shown. The principles in the design of this attachment are just opposite to those of the other, for in this case the roller *P* is located in the slotted plate and may be quickly removed through either of the end holes *R*, so that the side-head may be used for straight work during the same setting of the piece, without much trouble in preparation.

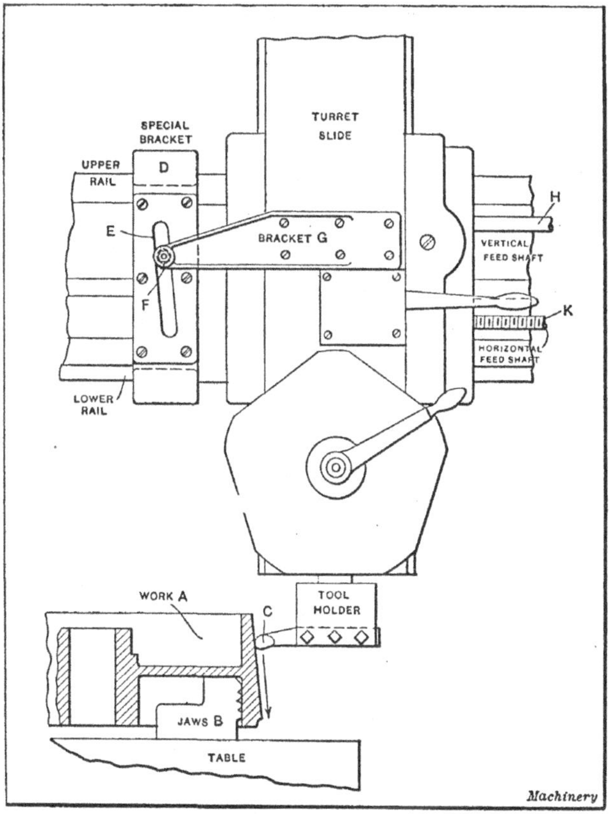

Fig. 16. Taper Attachment for a Vertical Boring Mill

The plate *H* is fastened to the rods F and *S* and vertical adjustment is obtained by sliding up or down. The binding screws *E* and *T* secure it in the desired vertical position. A T-slot *V* is cut along the entire length of the side-head slide and a cast adapting plate *O* is secured in it by means of T-bolts.

The angular plate *M* is screwed and doweled to the adapter. The brackets *K* and *L* are fastened to the side-head down slide and are fitted to the edges of the plate *H*, in order to avoid any bending action which might be produced by the pressure against the roller *P*. The brackets *D* and *Q* are similar to those used in the former case.

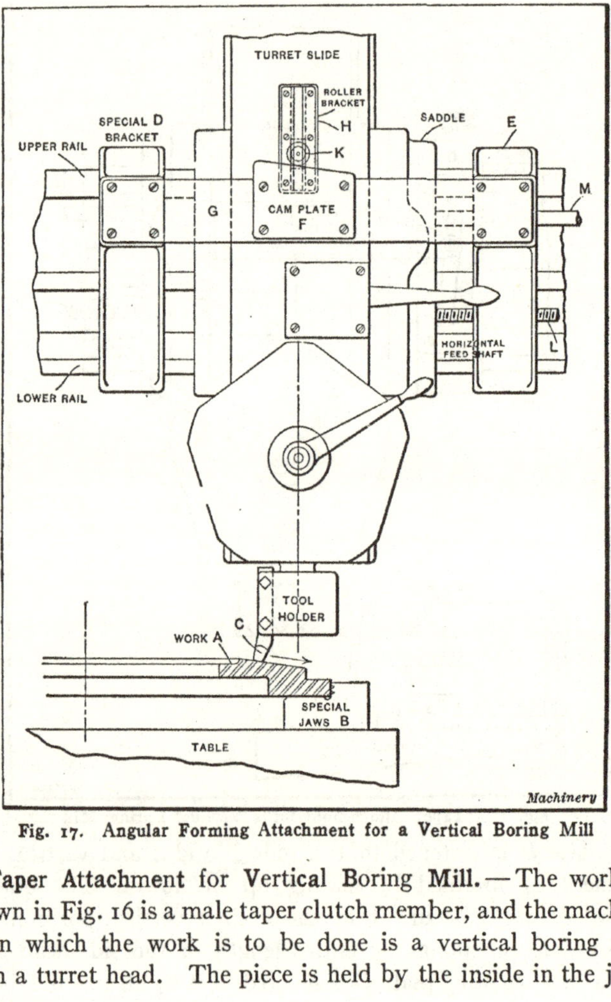

Fig. 17. Angular Forming Attachment for a Vertical Boring Mill

Taper Attachment for Vertical Boring Mill.—The work *A* shown in Fig. 16 is a male taper clutch member, and the machine upon which the work is to be done is a vertical boring mill with a turret head. The piece is held by the inside in the jaws

B and the tool *C* forms the taper. In order to permit lateral motion a special nut was required for the horizontal feed shaft *K*. This nut is not shown in the illustration, but was made somewhat on the principle of a lathe feed shaft nut so that it could be coupled and uncoupled rapidly. A special bracket *D* was fastened to the rail to the left of the turret slide and the forming plate *E* was fastened to it. A special bracket *G* was fastened to the turret slide and served as a support for the roller *F*. In use the vertical feed shaft *H* is thrown into gear and the turret allowed to float laterally as controlled by the forming plate *E*. When the other turret tools were to be used, the roller *F* was removed and the horizontal feed shaft nut recoupled. The action of this device was very satisfactory.

Angular Forming Attachment for a Vertical Boring Mill. — A very acute angle was to be produced on the work shown at *A* in Fig. 17 and a vertical boring mill was used to perform the operation shown. The work is held by the outside by the special jaws *B*; the tool *C* is used to perform the work. Two brackets *D* and *E* are bolted to the rails, one on each side of the turret slide and the cast-iron plate *G* is used to connect them and form a support for the cam plate *F*. A portion of the turret slide is machined off to permit the attachment of the roller bracket *H*. This bracket is slotted with a T-slot and the roller *K* mounted on a T-stud may be readily adjusted in it. In using this attachment it is only necessary to throw in the horizontal feed shaft gears and keep a downward pressure by hand on the cam plate F, by means of the handwheel on the end of the shaft *M*.

The various forms of taper attachments and devices which have been mentioned in this chapter cover nearly every variety of work and may be adapted to nearly any form of taper requirements that are met with in the course of general manufacturing.

CHAPTER IX

MACHINING CONVEX AND CONCAVE SURFACES

Machines for Spherical Turning and Boring. — The machining of convex and concave surfaces is a problem which frequently confronts the manufacturer, and its solution may be required under a great variety of conditions. The size of the work to be machined is a controlling factor, as it determines to a certain extent, the type of machine to be used. For example, the small steel cup washer shown in the upper part of Fig. 1 would naturally be machined on an engine lathe or turret lathe of the horizontal type, while the huge ball pipe joint in the lower part of the illustration would preferably be handled on some type of vertical boring machine. A manufacturing proposition may be necessary where one thousand or more pieces are to be handled, or it may be that only one piece is required. The work may be concave, or convex; in a plane perpendicular to the center of rotation, or it may be parallel with it and either internal or external. As the conditions governing the handling of work of this nature are so varied, and as the pieces themselves are of such widely different forms, it is very evident that we must consider several types of machines to which the forming devices may be attached. The construction of these devices must be adapted to the class of machine on which they are to be used, and this naturally influences the design of the attachments.

Important Points in Design. — The following rules should be considered when designing devices for machining convex and concave surfaces:

1. Whenever possible the attachment should be so designed that the form will be generated radially, so that the same portion of the tool will do the cutting at all times.

2. Try to use stock sizes of steel for the cutting tools so that replacements can be made easily.

3. See that the tool does not overhang too much, and that it is well supported and rigidly held. Care should also be taken that moving portions of the tool-holder or slide are of generous proportions and possess means of adjustment for wear.

4. Generate the curve by means of the machine alone, whenever possible, so that errors in the contour may not be occasioned by the failure of the operator to keep a certain roll or stud in contact with the forming plate.

5. When the attachment is of the type requiring the use of a roll and forming plate, it is essential to so arrange the plate that it will act as a guard against the tool being forced away from the work. That is, the action of the roll against the plate should be in the direction tending to carry the tool into the work, so that the thrust of the cut will always assist in keeping a positive contact between the roll and the plate. Counterweights or springs should also be used to obviate any tendency to draw in.

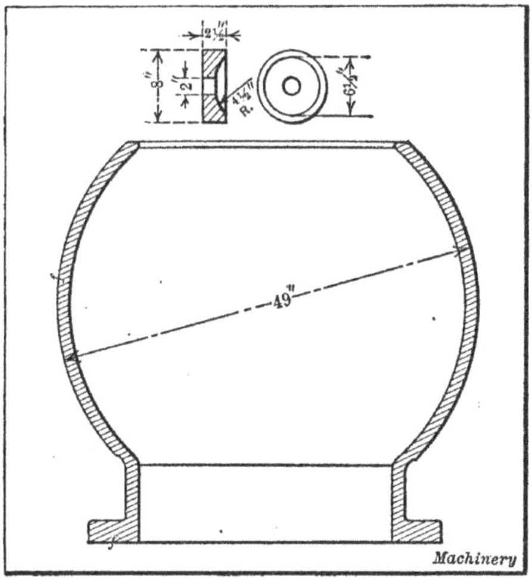

Fig. 1. Cup Washer and Ball Pipe Joint — Examples of Concave and Convex Turning

6. Economy in operating expense and the first cost of the attachment should also be considered, while the difference in workmen's rating is also a factor which should not be overlooked.

Radius Turning on the Engine Lathe. — When concave or convex turning or boring is required on only one or two pieces, or, in cases where it is not practicable to combine the radial work with other operations, the engine lathe may be adapted to a great variety of conditions. In manufacturing, it may occasionally be used to advantage, especially in cases when the length of time required to do work is sufficient to permit one man to run two machines.

Fig. 2 shows the simplest kind of a forming attachment for convex work, which is adapted to the engine lathe. The work *A* is held by the inside in the chuck jaws *M*. The bracket *H*

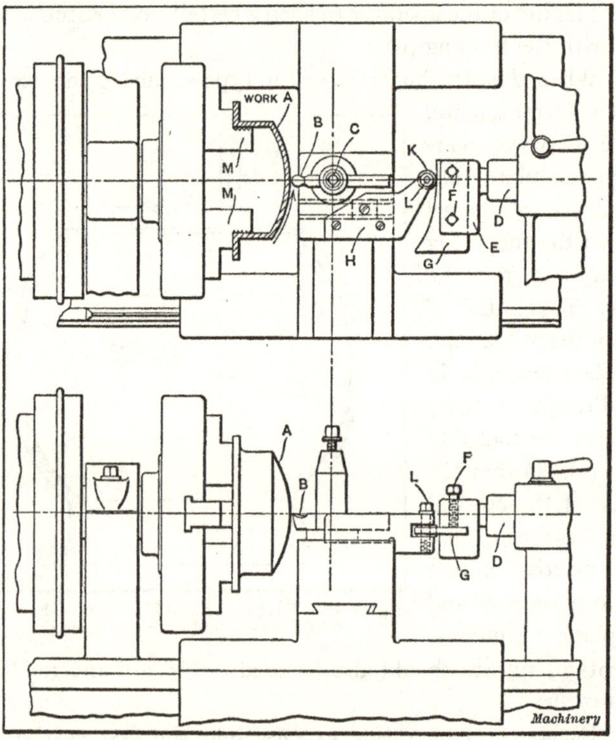

Fig. 2. Simple Attachment for Convex Turning in Engine Lathe

is screwed to the top of the cross-slide and carries at its outer end the tool-steel hardened and ground roller *K*, held in place by the screw *L*. The tailstock spindle *D* receives the holder *E* in which the plate *G* is inserted and secured by the two screws *F*. This plate is formed to the proper radius and is of tool steel unhardened. The cutting tool *B* is held in place in the regular toolpost *C*. The form of the cutting end of this tool is important as it must be formed to a perfect radius, in order that the cutting action may be uniform. In operating this attachment, the cross feed-screw is thrown into engagement, and the operator is required to force the roll *K* against the forming plate by means of the handwheel controlling the longitudinal feed of the carriage. An attachment of this sort requires the entire attention of the operator and,

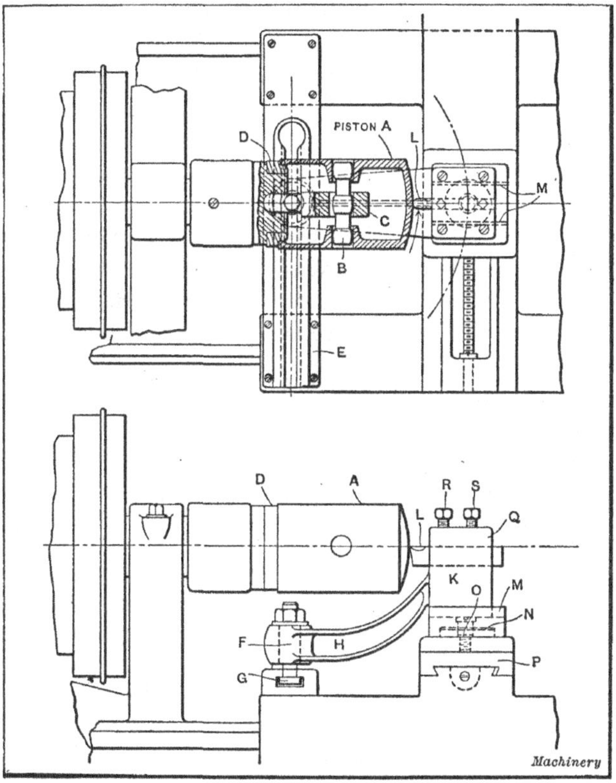

Fig. 3. Piston Crowning Attachment for the Lathe

therefore, variations are liable to occur in the contour of the work, due to imperfect contact between the roll and plate; hence it is a very poor attachment to use if there are many pieces to be machined.

Piston Crowning Attachment for the Lathe. — The arrangement shown in Fig. 3 was used for manufacturing in large quantities, and four lathes were equipped in this manner and required the services of two men to operate them. The piston *A* was located on a special draw-back chuck, on the steel locating ring *D*, and was drawn back firmly by the rod *C* acting on the pin *B* through the wrist-pin holes. A cast-iron bracket *E* is bolted onto the carriage of the lathe and is slotted at *G*. A cast-steel bracket *H* terminates at its upper end in the tool-block *K* which is dovetailed at *M* as shown in the plan view. The stud F is T-shaped at its lower end to fit the slot in the bracket *E*. A steel block is screwed to the top of the cross-slide *P*, and is shouldered at *N* to fit the upper block in which the dovetailed tool-block *K* slides. The screw *O* simply holds the units together. A steel plate *Q* contains the two screws *R* and *S* which securely hold the tool *L*. In operating this device, the cross feed-screw is simply thrown into engagement and its forward action causes the tool to swing radially at the desired distance from the center *F*, thus developing a spherical surface. This device was comparatively inexpensive and the results obtained by its use were very satisfactory.

Concave Turning with a Compound Rest. — A very simple device which may be used when one or two pieces only are required is shown in Fig. 4, but the radius which can be generated by this method is limited by the size of the compound rest. The work *A* is held by the outside in chuck jaws, and the tool *C* generates a radius equal to the distance from the end of the tool to the center of the swivel. The socket *G* is placed in the tailstock spindle and the overhanging end contains the round head pin *H*. The bar *D* is cup-shaped at *E* and *F* and bears against the end of the compound rest screw at *E*, while the other end engages the button at *F*. In using this arrangement, the holding-down gibs or straps of the compound rest swivel are set up to produce considerable friction, and the tailstock spindle is fed forward by hand, thus causing the compound rest to swing on its own center, thereby generating the desired radius. It is obvious that the carriage gibs must be tightened to prevent any longitudinal movement.

Radius-bar for Concave Turning. — A very simple arrangement for the engine lathe is shown in Fig. 5. The work *A* is held by the outside in chuck jaws and the round-nosed tool *B* is used to generate the concave surface. The tool is held in the ordinary manner in the toolpost *C* on the cross-slide of the lathe. A slotted holder *K* is tapered on its rear end to fit the tailstock spindle *L* and is slotted to receive the flat steel radius-bar *G* which is held in position by means of the shoulder screw *H*. The stud

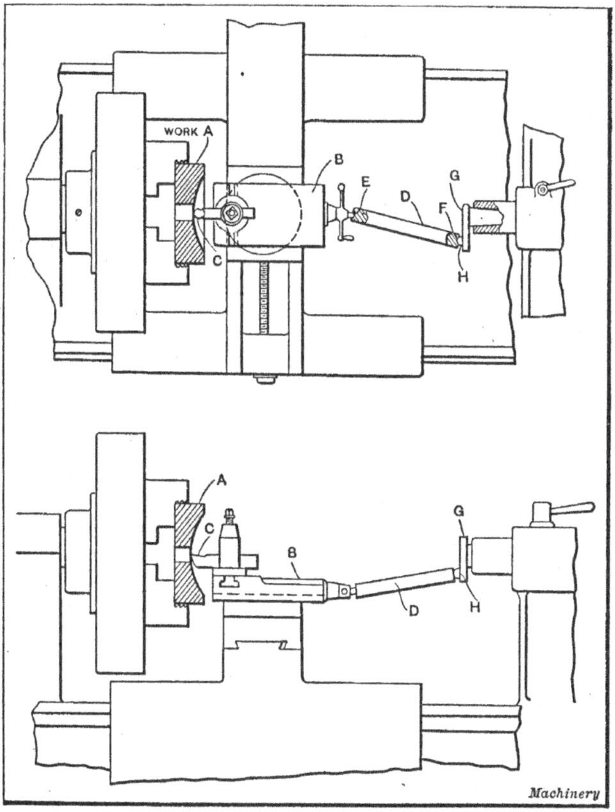

Fig. 4. Concave Turning with Compound Rest

F enters the tool slot of the cross-slide and serves as a pivot for the forward end of the radius-bar. This bar is made of the correct length to generate the desired radius. As in a previous instance, the cutting end of the tool *B* must be ground to a perfect radius. The arrangement shown here is a very simple one and may be used for various pieces of work by the substitution of a bar of the proper radius.

Pulley Crowning Attachment for the Engine Lathe. — Fig. 6 illustrates an arrangement that was applied to an old-style Pratt & Whitney lathe which had, at one time, been equipped with a taper attachment. The arrangement shown was especially

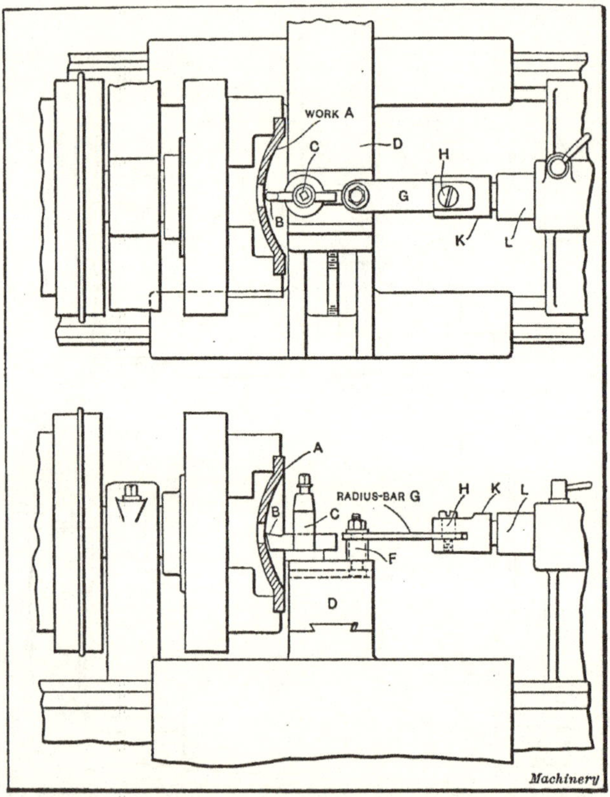

Fig. 5. Radius-bar Type of Concave Turning Attachment

designed for crowning pulleys which previously had been "chucked," faced and rough-turned straight on the periphery. A keyway had also been cut through the hub. A special arbor *D* was held on centers in a lathe, the dog *E* acting as a driver in the special faceplate F. The pulleys were put on the arbor until the face of the hub came up against the shoulder *H*, the spacer *J* being interposed between the two hubs. The cast-iron driving plate *K* was then put on, followed by the washer *M*. The nut *L* was then used to tighten the pieces in position.

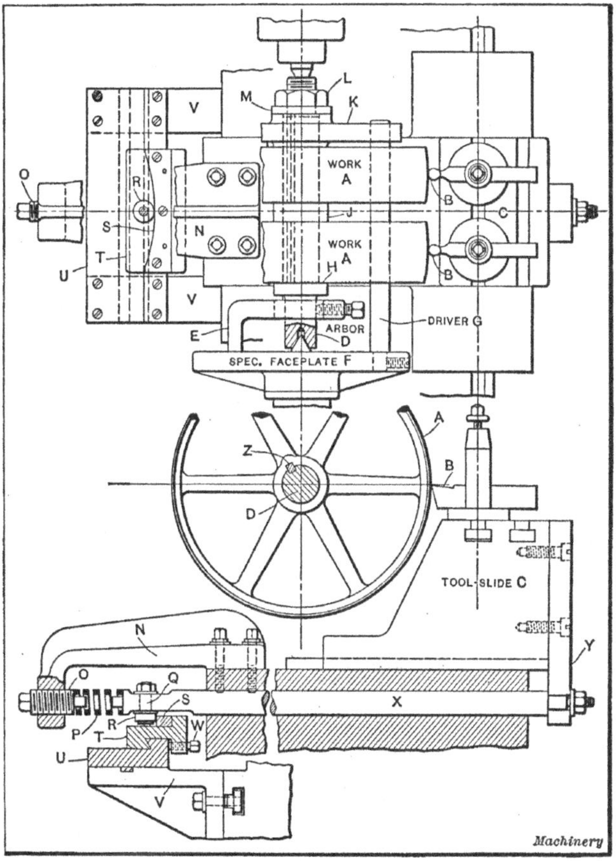

Fig. 6. Pulley Crowning Attachment for Engine Lathe

The driving bar *G* was used in order to prevent vibration which would otherwise be troublesome due to the thin flange of the pulleys. The two round-nosed tools *B* were mounted on the tool-slide *C* and held in the ordinary manner. A steel plate *Y* was bolted to the end of the slide and overhung sufficiently to permit the bar *X* to pass through it. This bar passed completely through the carriage and was a sliding fit in it. The stud *Q* passed through the bar and formed a bearing for the roller *R*. The cast-iron bracket *N* was fastened to the rear of the carriage and was tapped at its outer end to receive the screw *O*, which was

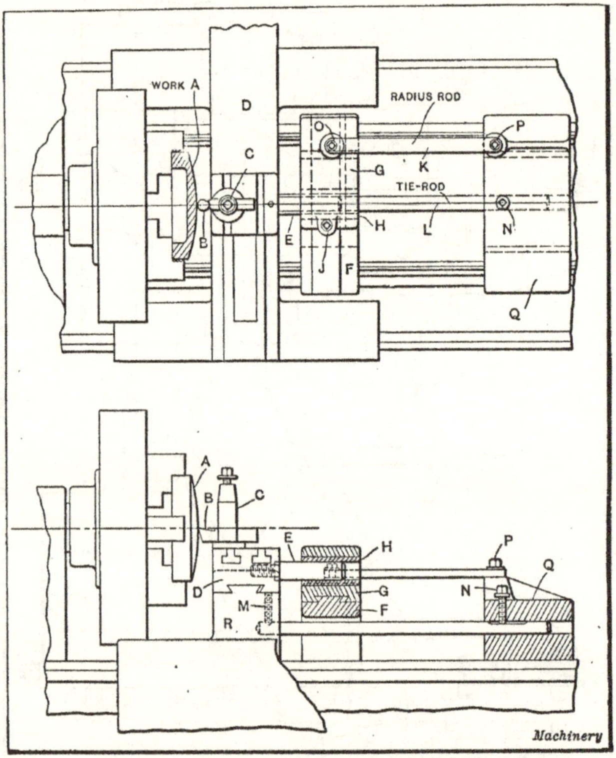

Fig. 7. Convex Turning Attachment of Radius-bar Type

used for adjusting the compression of the spring *P*. This spring was of square section ¼ inch in size and served to keep the roll in contact with the cam-plate *S*. Two brackets *V* and the plate *U* were a part of the taper attachment with which the lathe was originally equipped. The dovetail plate *T* served as an adjustable support for the cam-plate *S* and it was held in its proper location by the screws *W*. The upper view in the illustration is partially broken away to show the roll *R* in position against the cam-plate. The operation of this attachment is obvious and it is only necessary to state that its action was very satisfactory.

Convex Turning Attachment using a Radius-bar. — Fig. 7 shows a rather peculiar attachment for turning the convex surface on the cast-iron head-piece shown at *A*. The device is applied to the engine lathe, and its construction is such, that by the substitution of various lengths of radius-rods, it may be used for an infinite number of radii. The cutting tool *B* is held in the toolpost *C* in the ordinary manner and the longitudinal feed-screw is left free so that the carriage *R* may be perfectly independent. The bracket *F* is secured to the inner ways of the lathe so that it is absolutely prevented from moving. This bracket is dovetailed along its upper face and the slide *G* is mounted upon it. A tool-steel stud *E* is screwed into the side of the cross-slide and enters the bushing *H* which is contained in the bracket slide *G*. This bushing is eccentric and is split along one side thus permitting a slight adjustment to compensate for wear. The binding screw *J* holds it rigidly after setting.

The tie-rod *L* connects the carriage *R* with a special bracket *Q* at the rear end of the lathe, this bracket taking the place of the regular tailstock, and being gibbed to the ways in such a manner that it is free to slide longitudinally. The tie-rod is held in place by the screws *N* and *M*. The radius-rod *K* is made from a piece of flat steel and swings on the two screws *O* and *P*, located in the bracket slide and the tailstock substitute respectively. In the operation of this mechanism, the cross feed-screw is thrown into mesh and the slide *D* moves forward carrying with it the bracket slide *G* to which the radius-rod is attached. As the bracket *F* cannot move longitudinally, it is evident that the tail-stock substitute *Q* will be forced backward by the radius-rod and will carry with it the entire carriage and cross-slide, thereby generating the desired radius.

Ball Turning Device for the Horizontal Turret Lathe. — We now come to a somewhat more pretentious device designed for the horizontal turret lathe for the purpose of generating the spherical portion of the steel pipe joint shown at *A* in Fig. 8. This work is held in chuck jaws at *C*, the supplementary screws *B* being used in the outer ends of the jaws to assist in supporting the work. The regular turret lathe cross-slide and carriage are removed and the special slide *F* is substituted. It is gibbed firmly to the ways in the desired position by the gibs *G*.

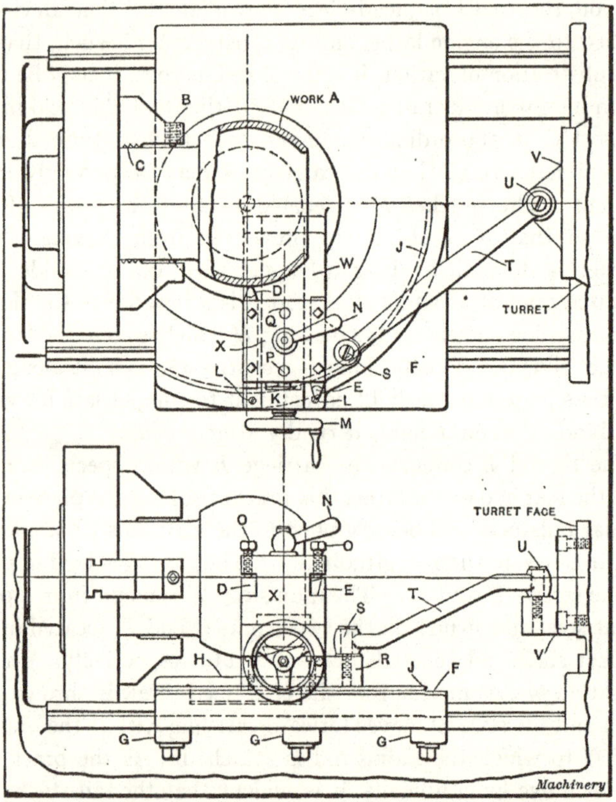

Fig. 8. Spherical Turning Device for Horizontal Turret Lathe

Directly under the center line of the spindle, a large circular recess is bored to receive the swivel *H*, and circular rim *J* is dovetailed so that the outer end of the swivel may be gibbed at *K* to insure rigidity. The two screws *L* hold the gib. The slide *W* is dovetailed and has an adjustment for diameters, controlled by the handwheel *M*. The roughing and finishing tools *D* and *E* are held in the indexing toolpost *X*, the tools being secured by the screws *O*. The index location is insured by means of the pin *P* entering a hole directly underneath it in the slide, and the other index position is determined by the hole *Q*. The binder *N* locks it rigidly. At the right of the swivel the lug *R* is built out to receive the pivot screw *S*, on which the forward end of the radius-arm *T* is fastened. A bracket *V* is bolted to the face of the turret and carries the screw *U* which supports the other end of the radius-arm.

In the operation of this device, the turret longitudinal feed is used while the roughing tool *D* removes the scale from the casting. The toolpost is then indexed and the finishing tool *E* completes the work. This attachment gave very good results and was satisfactory in every respect. There are several points in the construction of this device to which attention should be directed. One of the points is the turret toolpost by means of which the tools always remain set, so that the diameters are easily held to size. Another point is the dovetail gibbing of the outer portion of the swivel. This method does away with all possibility of chatter, providing the gib is kept tight. Another advantage is the adjustment for various diameters by means of the slide *W* which is mounted on the swivel.

Turret **Lathe** Attachment for **Crowning a Piston Head.** — The attachment in Fig. 9 is part of an equipment of tools for finishing the piston *A*, the work being held by the inside on a special chuck *B*, which is screwed onto the end of the spindle. There are turning tools fastened to the turret which are in action simultaneously with the attachment shown. These are not shown in the illustration, as they have nothing to do with the radius attachment. A special steel cross-slide *C* is mounted on the carriage in place of the regular slide, and the overhanging portion *G* carries a tool-block *M* in which the grooving tools for the piston are mounted. A special bracket *D* is firmly gibbed to the ways and secured by the screws *L*. The tool-steel forming plates *E* and F are screwed to this bracket and doweled in position. The plate *F* also acts as a strap for the overhanging portion of the cross-slide at *H*.

A hardened and ground tool-steel roller *J* is pivoted on the stud *K*, and controls the form of the crown by its contact with the plate *F*. The other plate *E* only acts as a guard. The

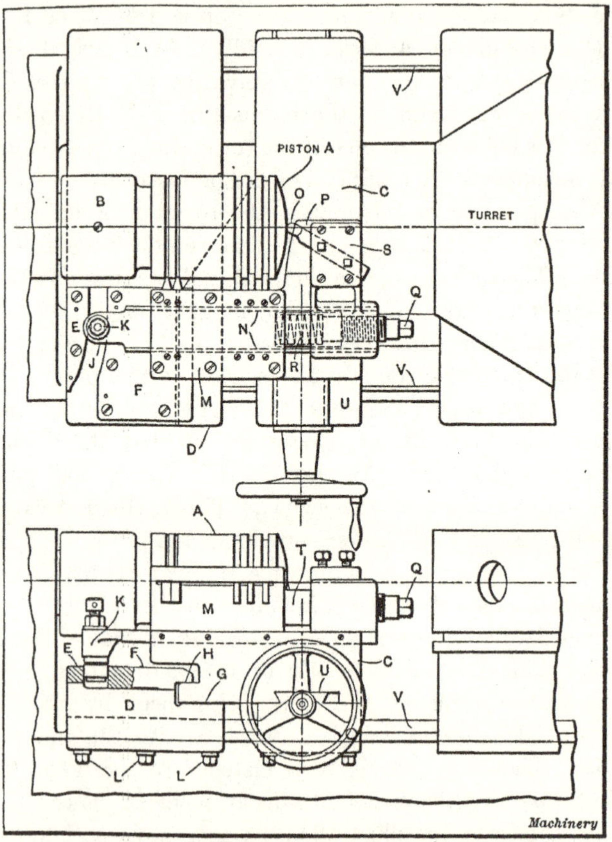

Fig. 9. Turret Lathe Piston Crowning Attachment

radius forming slide is dovetailed at *N* and passes through under the tool-block. It carries a tool *O* mounted in the tool-holder *S*. A supporting pad *P* is provided on the cross-slide directly under the tool-holder and serves to prevent vibration. A heavy coil spring *R* is used to insure proper contact of the roll with the cam-plate. The necessary adjustment is obtained by means of the screw *Q*. A brass tube *T* protects the spring from chips and dirt which might otherwise impair its action.

This attachment was built for a large manufacturing plant in the East, and proved very satisfactory. It is designed so

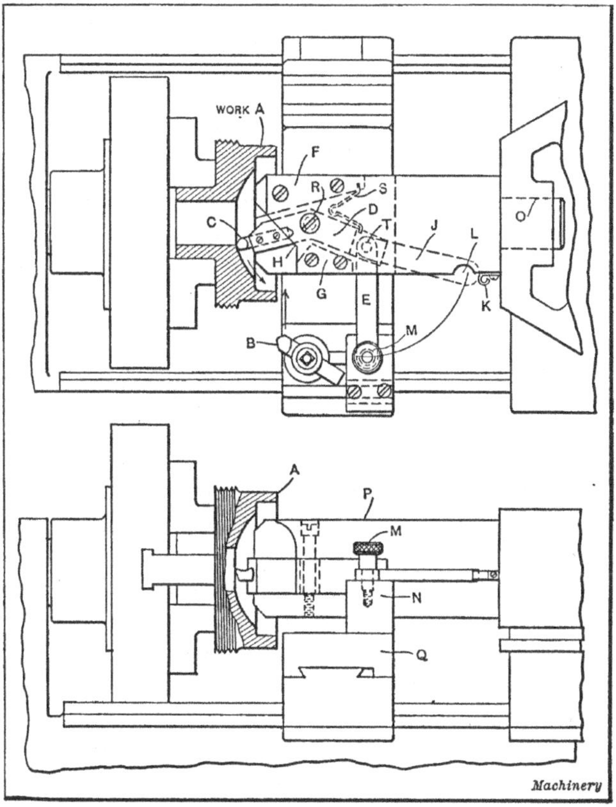

Fig. 10. Turret Lathe Attachment for Turning Concave Seat

that all the tools can work at the same time. In building the attachment, it was found necessary to fit the dovetail slides and other moving portions, after the tool-blocks were put in place, and the tools clamped in position. This was unavoidable because of the clamping strains, as these caused a certain amount of distortion, making it necessary to do the fitting after the tool-blocks were in position.

Radius Boring Attachment for a Horizontal Turret Lathe. — The device for generating the internal radial seat in the brass casting *A*, shown in Fig. 10, is somewhat out of the ordinary, and in addition to this, it is comparatively inexpensive. Furthermore, it is practically a self-contained unit, and requires no special fitting or attachments to the machine.

The steel bar *P* is held in the turret by the shank *O* and is secured in place by the turret binder. The bar is slotted completely through to receive the swivel arm *D* which is a machinery steel forging ground on two sides to fit the slot. The forward end of this swivel arm carries the tool *C* which is set at the proper distance from the screw *R* to produce the desired radius. At the end of the bar, a mill cut is made at *H* in order to permit access to the screws which hold the tool in position. The steel filler blocks *F* and *G* are put in to give additional rigidity to the end of the bar. The flat spring *S* is simply used to prevent back-lash in the swivel arm. The operating link *E* enters a slot in the end of the swivel arm and is held by the pin *T*. A special steel block *N* is fastened to the cross-slide and the knurled screw-pin *M* couples the end of the link to the slide. The tool *B* is used in connection with the attachment for facing the end of the work.

In operation, the cross-slide feed-screw is engaged and the radius nicely generated by the radial action of the arm. As soon as this operation has been completed, the knurled screw *M* is rapidly removed and the link *E* swung over into the slot *J* where it is held by the flat spring *K*. The cut *L* allows the fingers to grasp the link when the radial attachment is again put into use. This attachment has many good points, the only serious drawback being that there is a slight tendency to chatter when the cut is excessive.

Radius Boring-bar for the Vertical Turret Lathe. — The device shown in Fig. 11 was designed for use on the vertical turret lathe, and was used in connection with other tools in the main and side heads for the purpose of boring the radial portions of the automobile crank-case cover bracket shown at *A* in the illustration. The work is held on a locating fixture upon three pins *U* and is clamped by means of the straps *V*. Only a part of the work is shown, as the piece is a somewhat crooked one and the other portions have nothing to do with the radial boring. The fixture is held to the table by three T-bolts *W* and it is centered by the steel bushing *S*, which also acts as a guide for the stem of the boring-bar *R*. This bar

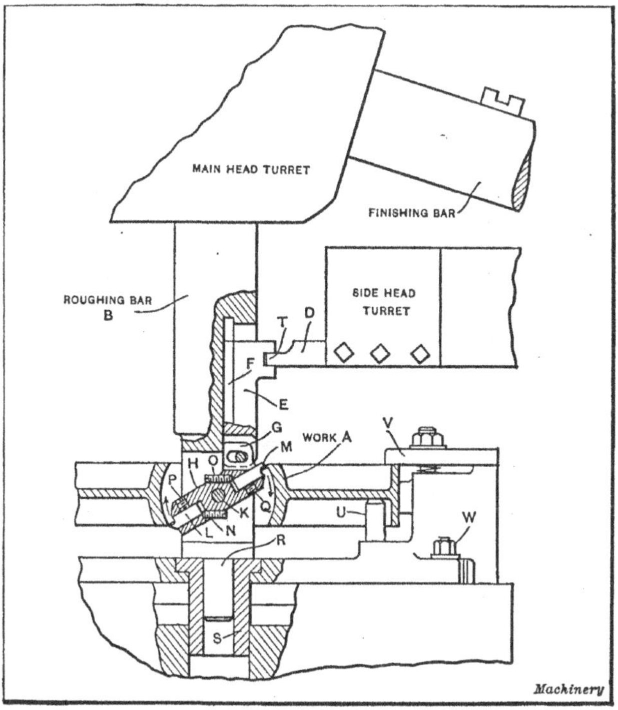

Fig. 11. Radius Boring-bar for Vertical Turret Lathe

is slotted out at its lower end to receive the swivel block *H*, which carries the tools *L* and *M* for rough-boring the radius. The block swings on the pin *K*, when forced to do so by the downward action of the operating arm *E*. The two cutting tools are backed up and adjusted by means of the screws *N* and *O*, and they are held firmly in position by the set-screws

P and *Q*. The bar *B* is slotted out to receive the tongue *F* on the operating arm, and the lower end of this arm contains a pin which works in a slot at *G* in the upper portion of the swivel tool-block. The steel piece *D* is held in the side-head turret and engages a groove in the upper part of the operating arm.

The finishing bar *C* is of the same general construction as the roughing bar except that only one tool is used for finishing, instead of the two shown in the roughing bar. It is well

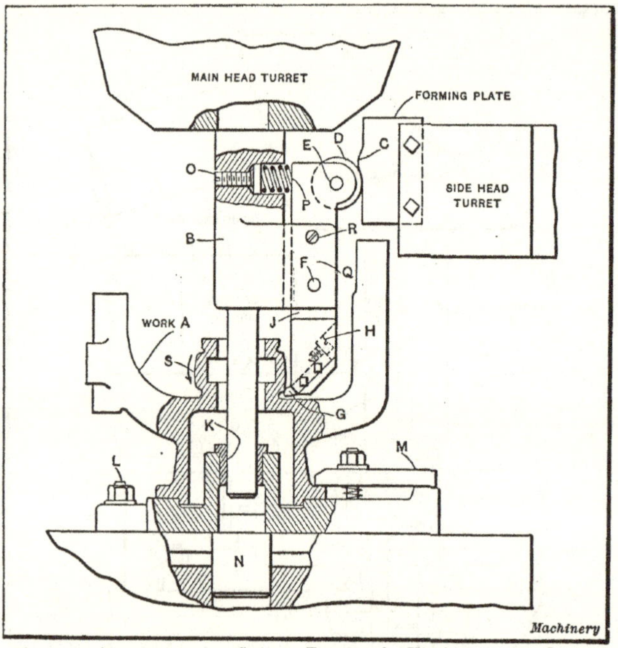

Fig. 12. Attachment for Convex Turning in Vertical Turret Lathe

to note that the use of two tools in roughing practically reduces the roughing time one-half, it being only necessary to cut half way with each tool, as one comes up from the bottom while the other is coming down, so that the two cuts meet at the center. In operation, the down feed of the side-head turret is started, thus providing the necessary power for operating the device. This scheme gave very satisfactory results as far as accuracy is concerned, but as the mechanism was confined in a small space, the size of the tools could not be made large enough to properly conduct away the heat generated in boring.

Convex Turning Device for a Difficult Piece of Work. — Fig. 12 shows a device for convex turning. The jack-shaft tube bracket *A* has been previously machined at its lower end, having been located for this setting by the finished surfaces. A special fixture is bolted to the table of the vertical turret lathe, by the three T-bolts *L*, and it is firmly secured by the straps *M*. The stud *N* serves to center the fixture on the table, while the bushing *K* acts as a guide for the end of the bar. The radial surface *S* is to be machined at this setting, and it will be noted that it is somewhat confined. The generating bar *B* is a 0.40 carbon steel forging, and the piloted end is carbonized, hardened and ground to a running fit in the bushing *K*. The projecting portion *Q* of the bar is of rectangular section and is slotted to receive the swinging arm *J*, which is pivoted on the pin F. This pin is equi-distant from the end of the cutting tool *G* and the center of the pin *E*. A hardened and ground tool-steel roller *D* revolves upon the latter, and is kept in contact with the forming plate *C* by means of the spring *P*, which is very stiff. Adjustment is provided through the screw *O*.

In operating this device, the turret down feed is used and the side-head turret is locked in the proper relation to the cutting tool, to produce the radius at the correct height on the casting. For the finishing cut, the roughing tool *G* is removed and a finishing tool substituted. A special screw *H*, having a large head against which the ends of the tools bear, makes this comparatively easy. The work can be turned closer to size if two roughing tools and a finisher are used, leaving only about 0.010 inch for the final cut. A defect in this fixture is that the thrust of the cut has a tendency to force the roll away from the cam-plate, and the action of the spring is sometimes insufficient to entirely overcome this. Had it been possible to design this attachment in such a way that the thrust of the cut would simply hold the roll more firmly against the forming plate, its action would have been more positive. The results obtained were sufficiently close, however, to conform to the required limits of accuracy, and it may therefore be stated that its work was satisfactory.

Convex Forming Attachment for the Vertical Turret Lathe. — A simple arrangement for the vertical turret lathe which permits the simultaneous use of both the main head and side-heads is shown in Fig. 13. The work *A* is a tractor pulley of large size. This is held by the inside, in the jaws *C* which are mounted on the raising blocks *D*. The boring-bar shown may be used

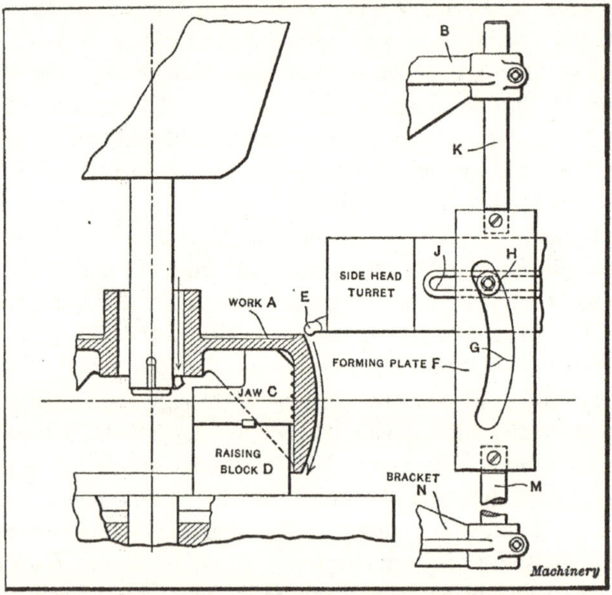

Fig. 13. Device for Crowning Pulleys in Vertical Turret Lathe

if desired, while the radius turning is taking place, as the radius attachment does not interfere in any way with the movements of the main head. The construction of this device is extremely simple and the results obtained by its use are very satisfactory. The upper and lower brackets *B* and *N* are attached to the bed of the machine and carry the rods *K* and *M*, which support the forming plate *F*. This plate is cut at *G* to the desired radius, and the tool-steel roller *H* travels along the slot and forces the tool *E*, which is held in the side-head turret, to take a similar path, thereby producing the convex surface on the rim of the pulley. A T-slot *J* is cut along the entire length of the side-head slide, and the roll *H* is fastened to a special bolt which can be adjusted to any position in the slot. The forming plate is also adjustable vertically, by sliding the rods up or down in the brackets.

Side-head Attachment using a Radius-bar. — Another method for crowning the outside of a tractor pulley is shown in Fig. 14,

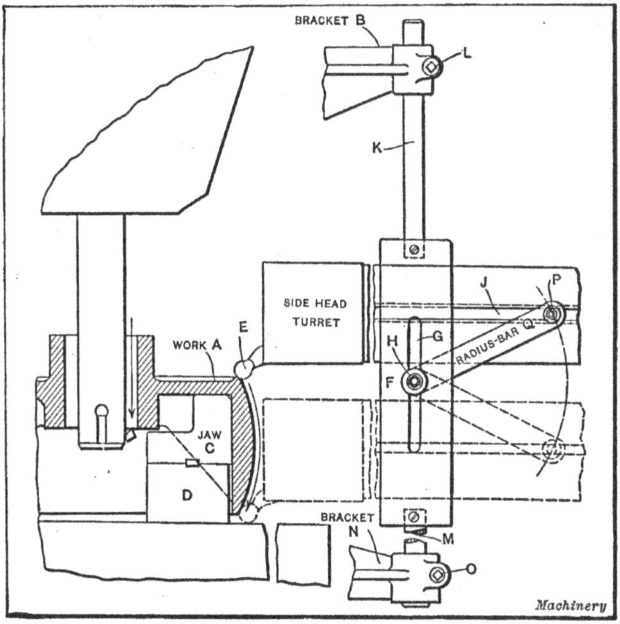

Fig. 14. Another Pulley Crowning Attachment

a radius-bar being used in this case to generate the desired curve. The work *A* is held, as in the previous instance, by the jaws *C* on the raising blocks *D*. The same brackets are also employed, but the plate F is not used for forming. It is slotted at *G* for adjustment only and a special screw or stud *H* is used in the slot as a pivot for the radius-bar *Q*. A T-slot *J* is cut along the entire length of the side-head slide and receives a special stud *P*, to which the other end of the radius-bar is fastened. The down feed of the side-head is thrown into engagement when operating the device, and the tool *E* naturally follows the radial path controlled by the length of the radius-bar. This device is simple and good and only requires extra radius-bars in order to handle a great variety of work.

Vertical Boring Mill Attachment for Spherical Turning. — The male member of a very large ball-and-socket pipe joint is shown at *A* in Fig. 15. The spherical portion, which is to be machined, has an outside diameter of forty-nine inches.

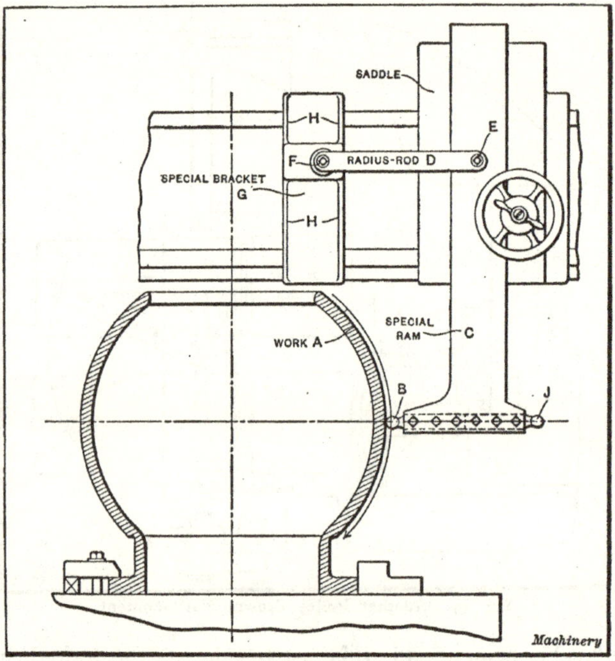

Fig. 15. Spherical Turning Attachment for Vertical Boring Mill

This member fits the female portion shown in Fig. 16, and the attachments used for boring and turning are the same, although the locations and the tools are different. A special bracket *G* is mounted on the rail and is gibbed at the rear so that it can be adjusted easily. It is well gibbed at *H* to insure rigidity. The special ram *C* was made in the form shown, its lower end acting as a tool-holder for the tools *B* and *J*; the former is used for the outside turning and the latter for the inside boring. The ram itself is a steel casting of extra-heavy section on account of the excessive overhang required. The radius-rod *D* was fastened at the two ends by the screws F and *E*, and was made of the proper length to give the correct radius. A special arrangement, not shown in the illustration, permitted a side floating action to the saddle along the rail, when the down feed was engaged.

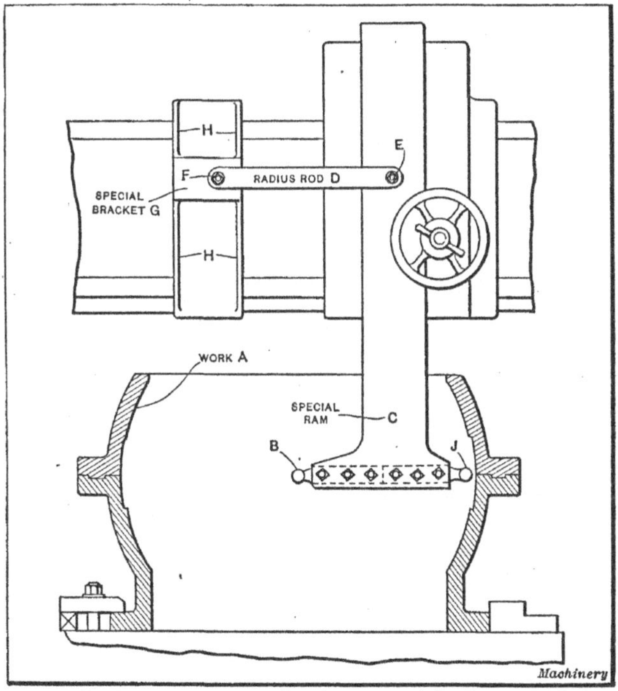

Fig. 16. Attachment shown in Fig. 15, used for Internal Work

For handling the socket portion of the joint shown (Fig. 16), the entire mechanism was moved over on the rail far enough to bring the tools in the desired position for boring, the tool *J* being used for this purpose.

Attachment for Convex Turning in a Horizontal Plane. — The work *A* shown in Fig. 17 is a steel casting which has been previously machined at *B*. It is held on a fixture (located centrally on the table by the plug *C*) and clamped down by the three clamps around the flange. Two brackets *G* and *H* are mounted on the rail and serve to carry the supporting plate *J*. As these brackets necessarily extend some distance in front of the rail they are strongly ribbed, as shown in the illustration.

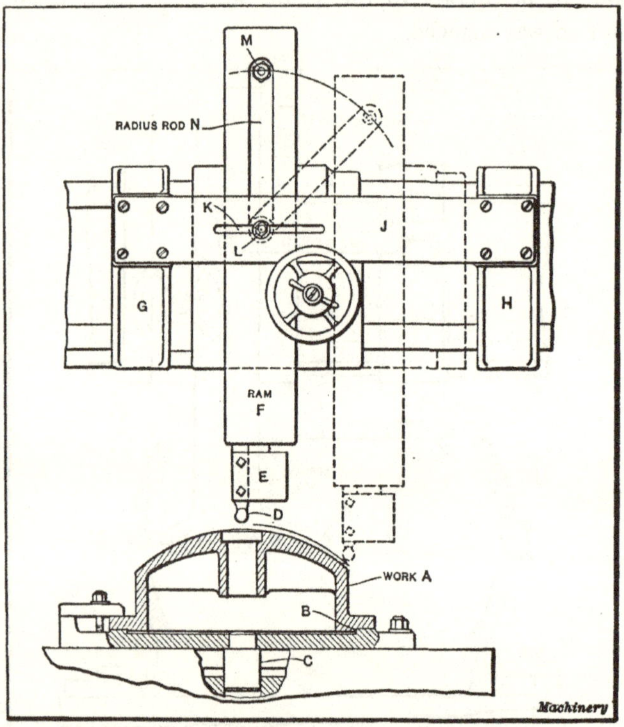

Fig. 17. Convex Turning in a Horizontal Plane

The supporting plate *J* has a slot cut in it at *K* for adjusting the pivot stud *L* in a longitudinal direction. This stud supports one end of the radius-rod *N*, and the other end is attached to the upper portion of the ram *F* by pivot stud *M*. A regular tool-holder at the lower end of the ram holds the tool *D*. No special arrangement is necessary to allow the saddle to "float" in a transverse direction, as the horizontal feed-screw is used in this case, the vertical feed being simply thrown out of mesh. This permits the ram to float vertically, as controlled by the radius-rod *N*.

Attachment for Concave Turning in a Horizontal Plane. — The steel casting shown at *A* in Fig. 18 is turned concave by the same device as that illustrated in Fig. 17, the only difference

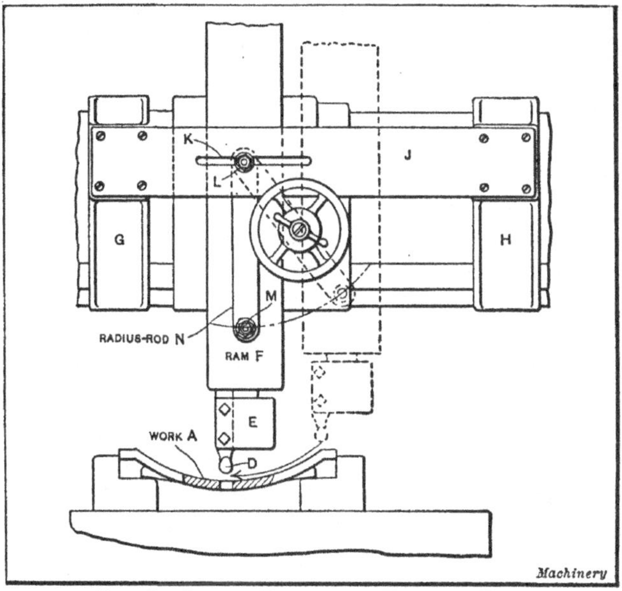

Fig. 18. Concave Turning in a Horizontal Plane

being that the radius-rod *N* is pivoted to the ram below the supporting plate instead of above. The operation of the mechanism is exactly the same as that for the convex surface.

Many variations of the devices described in this chapter have been used for generating radial surfaces, but enough have been described to enable the reader to select a method most suited to the particular problem that may require a solution. Adaptations may be readily made to fit almost any condition likely to be encountered in general manufacturing.

CHAPTER X

METHODS FOR MACHINING ECCENTRIC WORK

Factors which Determine Methods to be Used. — Castings or forgings having two or more cylindrical surfaces eccentric to each other may be machined by a variety of methods, that most suitable for any particular case being dependent upon a number of variable factors. In order to handle the work to the best advantage, these factors must be carefully considered. In the first place the machine most suited to the work must be selected, and this point is partially dependent on the size and shape of the work itself. The number of pieces to be machined is an important factor, in that it has an influence on the permissible tool outlay, for it is apparent that a high tool cost would not be expedient in the case of a lot of 200 pieces; whereas, if 5000 pieces were to be machined the cost of tooling and fixtures could almost be neglected provided substantial gains in production were to be made by some improved method of handling. There are cases when the work can more profitably be handled in two settings by means of a fixture, while in other instances the use of an eccentric turning device seems to offer the best solution of the problem. In the first case, a fixture of some sort is required and this usually presents no serious difficulties to the designer unless the work is very complicated. However, if an eccentric turning device seems warranted (as might be considered advisable if a great number of pieces were to be machined), it would then be necessary to design something on this order which would produce the work in the shortest possible time.

In the design of fixtures and devices for eccentric work there are a few points which require most careful consideration. Let us first give our attention to those which relate to special fixtures, for these points may be quickly summarized: If the work is to be finished in one setting, then some form of indexing fixture must be designed which will give the correct amount of " throw " when it is set over, and a great deal of care is required so that no change in the clamping of the work will be necessary between the first and second settings of the fixture. Indexing points, bushings and pins should be protected from dirt and chips so that inaccuracies will not occur due to imperfect contact at the locating points. When the work requires two settings the piece must be carefully located from a previously machined surface which bears a positive and fixed relation to the eccentric portion. The clamping must be rapid and must be so arranged that it will not distort the work.

Points in Design of Eccentric Turning Devices. — When it is desired to use an eccentric turning device, the following points in design will be found of value:

1. Rigidity. All brackets and castings should be made of ample section not only to withstand the stresses to which they may be subjected but also to deaden the vibration which may be caused by cutting action of the tool. A bracket may be strong enough to perform its functions and yet not have sufficient metal in it to prevent vibration. Proportions of moving shafts should be generous and bearing surfaces of good length.

2. Direction of the cam thrust. This should be such that it tends to throw the tool into the work rather than in a contrary direction. Springs or weights can be arranged to prevent the tool from "digging in." This arrangement is desirable but not absolutely essential, for conditions are sometimes such that the cam must be arranged so that it throws the tool away from the work. When it is found necessary to make it in this way the springs or weights should be made very stiff so that a positive contact with the cam is assured at all times.

3. Proportions of cam roll. In cases where a rolling contact is used, the pin on which the roll revolves should be not more than one third the diameter of the roll. If not so proportioned the friction on the pin, caused by the pressure of the contact with the cam, will be so great that the roll may not always

revolve and this will naturally have a tendency to make flat spots in the roll due to unequal wear.

4. *Chip protection.* All moving parts should be carefully guarded so that chips and dirt will not get in and cause imperfect work.

5. Provision for oil. This matter should be taken care of on the eccentric turning device just as if it were a part of the machine itself, for many a good attachment has been seriously injured by lack of oil. Oil cups are cheap and easily applied and an operator has no excuse for neglecting to oil a fixture when oil cups are in plain sight.

6. Simplicity and economy in cost of attachment. These points should receive due consideration, always bearing in mind the number of pieces to be machined and making the cost in proportion as far as possible.

A number of illustrations of eccentric work are shown, and will be discussed and criticised in this chapter.

Eccentric Turning in the Engine Lathe. — The upper illustration in Fig. 1 shows the simplest problem in eccentric turning which may be met with in ordinary shop practice. The steel eccentric shaft *A* is to be machined so that the smaller end *J* will be eccentric to the larger portion by half an inch. The centers *B* and *C* are first carefully put in at the correct distance from each other, after which the work is placed in the lathe (on the centers *B*) and the entire shaft turned concentric. In turning the portion *J*, the centers *C* are used, the dog *D* being used as the driver. The work revolves on the centers *E* and *F* in the spindle and tailstock of the lathe, and the tool *G* in the cross-slide toolpost *H* is used to do the turning. It is obvious that several cuts are necessary to bring the work down to the proper size. Work which requires machining in this manner is not at all unusual, and this method of handling is familiar to every machinist and toolmaker.

Sometimes work which requires eccentric turning is of such a nature that the use of centers is not desirable.. Such an instance is shown at *K* in the lower part of Fig. 1. In this case the flange and the hole *P* are concentric, but the hub *O* is $\frac{3}{8}$ inch eccentric. There are several ways in which this piece of work could be handled: It could first be held in a four-jawed chuck by the hub *O*, the jaws being thrown off center until the flange runs approximately true. The flange could then be turned and faced and the hole *P* chucked out and reamed. An eccentric arbor could then be used in the hole and the hub *O* turned on this arbor. This method would require a special arbor with

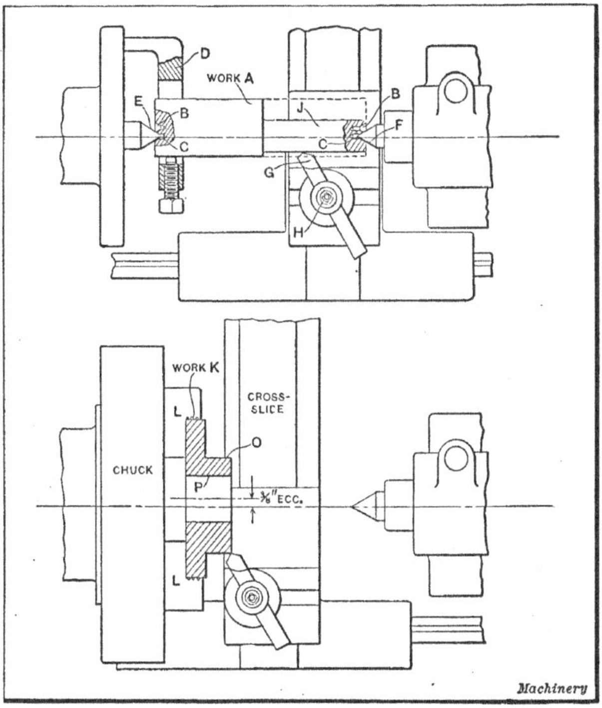

Fig. 1. Eccentric Turning in the Lathe

the proper eccentricity. In order to avoid this expense let us consider that we have performed the first operation as mentioned and are now ready for the eccentric turning. By using a four-jawed chuck and holding the work in the jaws *L* by the flange it is a very simple matter to indicate an arbor or stud placed in the hole *P* and set the jaws off center until the desired eccentricity has been obtained.

Eccentric Turning Attachment for the Engine Lathe. — The two methods given in Fig. 1 are more suited to work not made in quantities, and may be used in cases where one or two pieces are required. The device shown in Fig. 2 was used for a number of pieces, there being fifty in the lot. This arrangement was applied to an old Pratt & Whitney lathe which had a taper turning attachment at the rear of the bed, this machine being selected because it could be adapted quickly to the work in hand without great expense. This example is given not be-

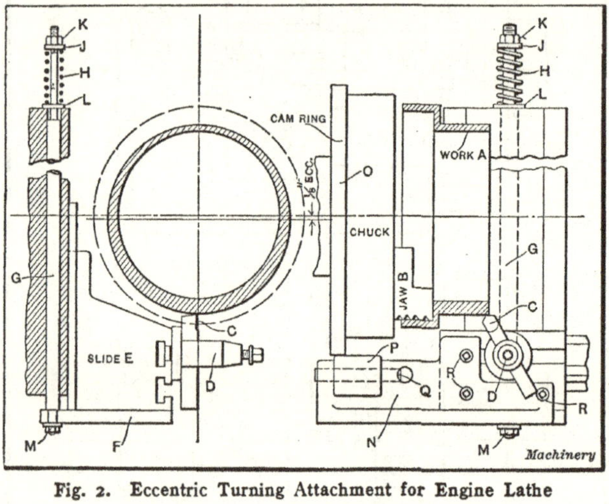

Fig. 2. Eccentric Turning Attachment for Engine Lathe

cause it is of great value but simply to call attention to a "make-shift" method of handling a job which was "rush" and which, therefore, had to be completed in the shortest possible time. One of the hubs on the work *A* was eccentric to the other by $\frac{1}{8}$ inch and the work was held by the inside in the jaws *B*. A cam ring *O* was set eccentric and fastened to the back of the chuck by four screws not shown. A cast-iron bracket *N* was screwed to the top of the cross-slide *E* by the three screws *R*, these screws entering shoes in the T-slots. A hardened steel roller *P* revolves on the pin *Q* at the outer end of the bracket. A steel plate *F* fastened to the front of the cross-slide overhangs the end and acts as a support for the rod *G*, being held in place by the nut and washer at *M*. This rod runs entirely through the carriage and is threaded at its outer end to receive the nut *K*.

A stiff coil spring *H* bears against the two washers *J* and *L*, and serves to keep the slide drawn forward at all times so

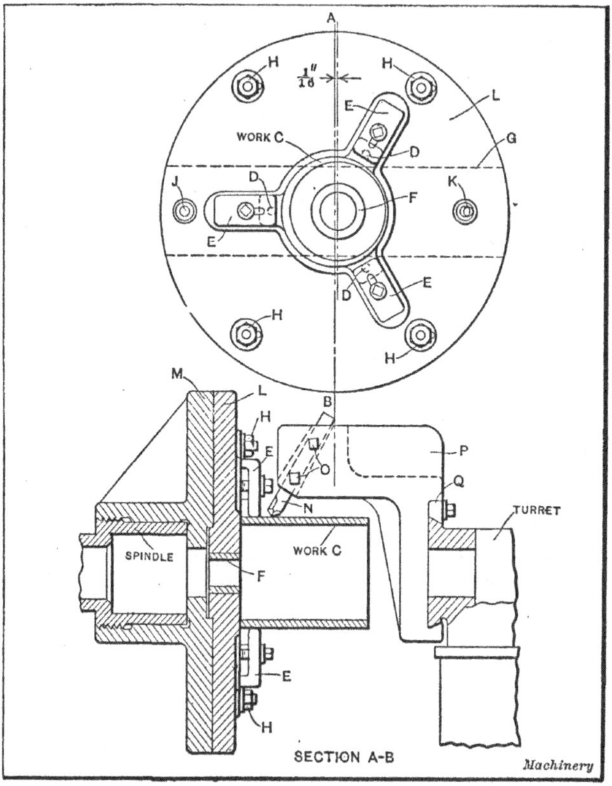

Fig. 3. Eccentric Turning Fixture for Horizontal Turret Lathe

that the roll *P* is continually in contact with the cam ring *O*. A tool *C* held in the toolpost *D* serves to turn the eccentric. It will be noted that the cam roll *P* is made long enough to give contact during the required cut.

Eccentric Turning Fixture for a Horizontal Turret Lathe. — Packing rings such as are used on automobile gas engines are usually eccentric and these are handled in many cases by an automatic or semi-automatic machine which is provided with some sort of an eccentric turning device. There are instances, however, when the expense of a device of this kind is not warranted by the amount of production required. Fig. 3 is an example of this sort, the rings being required in small quantities. The ring pot *C* is of cast iron and has been previously faced on the back of the lugs, approximately square with the pot proper. Holes have also been drilled at *D* by the use of a jig. The work is located by these three holes on the pins *D* which are fixed in the sliding plate *L*. The three clamps *E* are used to hold the pot firmly, and these clamps are slotted to facilitate rapid removal. The bushing *F* is concentric with the interior of the pot and is used in another operation as a guide for the pilot of the boring-bar. The plate *L* is tongued at *G* to fit a corresponding groove in the faceplate *M*. This faceplate is well ribbed and is screwed to the spindle. After the boring has been done, the four nuts *H* are loosened, the locating pin *J* is removed from its bushing and the plate is pushed over until the pin *J* can be pushed " home " in the other bushing *K*. It will be noted that it is not possible to locate the two indexing bushings side by side on account of the small amount of eccentricity, this being only $\frac{1}{16}$ inch in this case.

After the plate has been set over, the nuts *H* are tightened and the turning tool *N* is used to turn the eccentric surface. This tool *N* is firmly secured in the body *P* by the two set-screws *O* and it should be noted that the slot in which it rests is tapered toward the lower end so that a swinging adjustment may be made. The body of tool *P* is secured to the dovetailed face of the turret by gib *Q*. This fixture, although somewhat slow, gave excellent results.

Eccentric Ring Turning Device for a Horizontal Turret Lathe. — The device shown in Fig. 4 was designed and manufactured by the Pratt & Whitney Co. for use on its horizontal turret lathe, and it has been remarkably successful in the production of eccentric piston rings for gas engine work in a minimum amount of time. The construction of the entire device is such that it may be applied to a standard machine without injuring it for other work. The component parts of the device can be removed quickly from the turret lathe, leaving the machine ready for regular chucking work. A special faceplate *H* is used to hold a three-jawed geared scroll chuck having special jaws *B* which grip the angular flange of the ring pot. This flange is made so that it is concentric with the interior of the pot. The faceplate *H* is furnished at its rear end with a spur gear *J* which meshes with another of the same size *K*. The cast-iron bracket F is fastened to the top of the spindle cap and has a hub *G* at its outer end which acts as a bearing for the spline shaft *L*. It will be noted that the gear *K* is pinned to this shaft by a tapered pin. A large cast-iron bracket *U* is fitted to the dovetailed faces of the turret and extends entirely across it, being secured by the gibs shown. A boring-bar of tool steel, having a hardened and ground pilot *D*, is mounted in the turret, and the pilot is guided by a bushing *C* in the chuck. The boring tool *E* sets at an angle of 30 degrees in the bar and is secured by two square-head set-screws. Fine adjustments are obtained by means of the backing-up screw shown. The piloting of the bar in the chuck bushing gives great rigidity.

A steel block *S* holds the tool *R* which does the eccentric turning. This block is mounted on a dovetail slide *T* which has an adjustable taper gib securing it to the outer end of the bracket *U*. A small L-shaped bracket *O* is tongued to the top of a lug on the main bracket and acts as an outboard bearing for the cam quill *N*. This quill has a roll with the correct eccentric turned on it and it is supported at each end in two hardened and ground steel bushings, these bushings being forced into their places in the bracket and lug, respectively. A smaller roll *Q* is mounted on the top of the dovetailed slide and is held constantly in contact with the cam roll by the action of an adjustable spring of heavy section, concealed in the body of the slide and adjusted by means of a screw. (To avoid complicating the drawing this spring is not shown in the illustration.) The cam quill is furnished with a long key which passes completely through it, this key engaging and being free to slide in the spline *M* in the shaft *L*. A tool-block *V* of solid steel

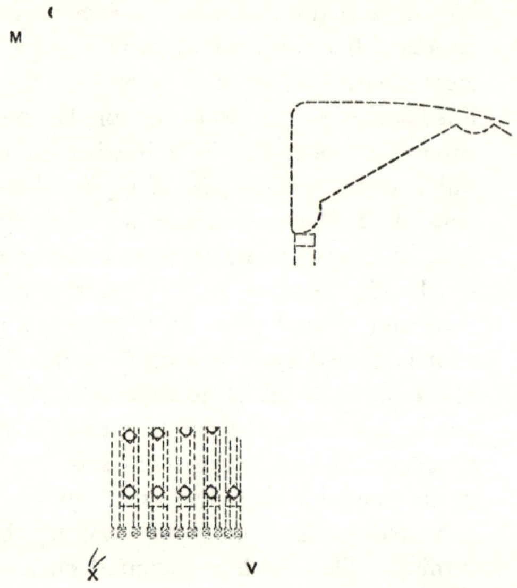

Fig. 4. Eccentric Piston Ring Turning and Boring Attachment

is mounted on the front of the cut-off slide, and attention is called to its construction. The two side pieces are tongued at *Y* and are screwed fast to the top and bottom pieces of the tool-block, the tongues serving to take the thrust of the holding-down screws and making for a very rigid construction. A supplementary block is fitted so that it slides freely in the holder and in this block the cutting-off tools *X* are mounted. Fine adjustment is provided in the backing-up screws. The tools themselves are clamped in position by the blocks *W* which bear along their upper edges and are secured by means of the square-head set-screws in the tool-block cap, two tools being gripped by each block. The cutting ends of these tools are ground to an angle of 5 degrees with the center line of the spindle so that their action in cutting off will be progressive, and the rings will drop off one at a time without breaking. In the spacing of the cutting-off tools an allowance of 0.010 to 0.012 inch is made for grinding the sides of the rings. These tools are very carefully made and surface-ground on the side so that they are a nice push fit in the tool-block slots. Oil cups for the various bearings are shown at *Z*.

The operation of the device is as follows:- As the spindle revolves the spur gear *J* meshing with *K* revolves the spline shaft *L* once while the spindle revolves once, and the quill which is keyed to a sliding fit on the spline shaft revolves the cam roll *N*, thus forcing the slide *T* in and out and forming the eccentric on the outside of the pot. As the turret moves forward, it will be seen that the boring tool *E* is working on the concentric inside while the tool *R* is in opposition to it on the outside. After the turning and boring have progressed about an inch and a half, the cross-slide feed is thrown into engagement and the cutting-off tools part the rings one by one. A great number of these equipments have been sold by the Pratt & Whitney Co., and have given universal satisfaction.

Eccentric Turning Device for a Cast-iron Eccentric. — The cast-iron eccentric shown at *A* in Fig. 5 has the flange *B* and the hole *C* concentric, while the portion *A* is one half inch cccentric to these surfaces. The work is held by the inside of the

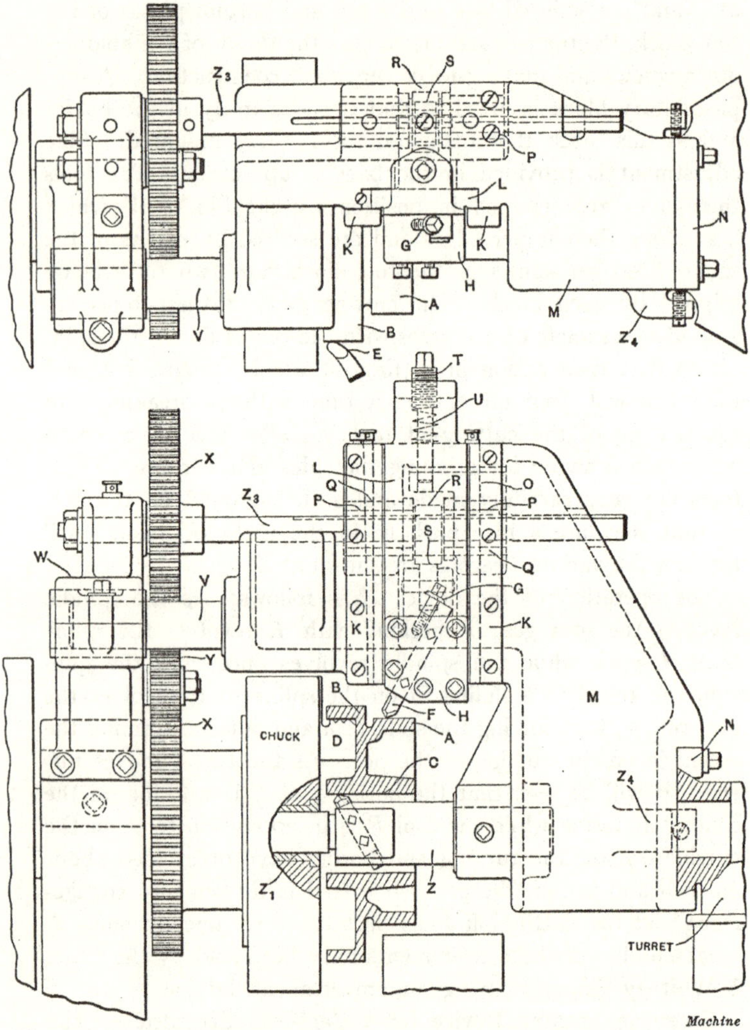

Machine

Fig. 5. Eccentric Turning Device for Cast-iron Eccentric

flange in the special jaws *D*, a three-jawed geared scroll chuck being used. The faceplate of this chuck has a spur gear *X* cut on its rear end. This gear meshes with an idler *Y* which, in turn, transfers the motion to the upper gear *X*, this being of the same diameter as the faceplate gear. The gear *Y* is mounted on a stud which has a bearing in the bracket *W*, this bracket being mounted on the spindle cap. The shaft Z_3 on which the upper gear *X* is mounted, is splined to receive a key which extends the entire length of the cam quill *P*. This key is fitted so that it has an easy sliding fit in the shaft. The bracket *M* is of cast iron, box section, and it is mounted on the dovetail face of the turret and secured by the gib *N*. The two lugs Z_4 each contain a set-screw which is used partly to adjust the heavy bracket into position when assembling, and partly to prevent its working loose on account of the jar incident to the shock caused by indexing the turret. The piloted boring-bar *Z* is mounted in a boss in the bracket and is used to bore the hole in the hub while the eccentric turning is being done. The bushing Z_1 in the chuck acts as a guide for the bar and also assists in preventing chatter. It is well to note that the boring tool stands in a vertical plane. The bracket has a pilot *V* at its forward end and this enters a bushing in the spindle cap bracket. The hub or boss which holds it is split and is clamped tightly on the bushing by means of the collar-head screw shown. The tool *F* which turns the eccentric is mounted in a steel block *H* and it is backed up by the screw *G*.

After the rough-turning operation is over, the tool F is removed and a finishing tool substituted for it, the backing-up screw acting as a gage or stop against which the ends of the tools rest so that diameters can be duplicated without trouble. The boring-bar also can be quickly removed and the finishing bar substituted for it by simply loosening the screw which holds it in place in the hub. The tool-block is secured to the slide *L* by the screws shown and the slide is held in place by the two strap gibs *K*. A taper gib with a backing-up screw is provided to take up wear on the slide. A projecting lug at the top of the slide provides a means of holding the adjusting screw *T* which is used to stiffen the heavy coil spring which holds up the slide. The bracket *O* which furnishes one of the bearings for the cam quill is tongued on its upper surface and is screwed fast to a projecting portion of the main bracket. The two ends *P* of the cam quill are carried in bushings *Q*. The cam roll *R* is ground to the correct eccentricity and controls the movement of the slide through the contact of the roller *S* which is mounted between the lugs shown on the back of the slide. A cored opening in the main bracket permits the free passage of the lugs without interference.

During the operation of the eccentric turning device and the boring of the hole (power feed being thrown in for this work), the tool *E* which is held in a regular toolpost on the front of the cut-off slide is fed by hand along the concentric portion of the casting *B* and is also used for facing the end. While the finish-turning and boring are being accomplished by the eccentric device, a finishing tool on the rear of the cut-off slide is used to finish the surface *B* (this tool is not shown). Oilers are provided for all bearings. It may be noted that the operations of turning eccentric and concentric and boring the concentric hole are all going on at the same time. The device is somewhat top-heavy and rather expensive, but it is capable of handling a number of pieces of varying eccentricity by substituting cam quills of the proper "throw," and making the necessary boring-bar adjustments. The coil spring *U* is of $\frac{3}{16}$-inch stock, square section, and is very stiff, as it has to support the weight of the slide and keep the tool from pulling in. Attention is called to the fact that the action of the cam roll is such that it forces the tool to take the correct path so that an absolutely correct eccentric is formed.

Machining an Eccentric Hub on the Vertical Turret Lathe. — The work shown at *A* in Fig. 6 has a hub *B* which is eccentric to the gear *C* on the outer rim by $1\frac{1}{2}$ inch. In the machining of this casting it was first held by the inside of the flange in a three-jawed chuck having special jaws permitting the edges of the flange to be faced. In the setting shown in the illustration, a four-jawed chuck was utilized, the jaws *D* being set off center sufficiently to produce the desired eccentric on the hub. As there were only a few of these pieces to be machined the expense of an eccentric turning device was not warranted, so that this method, although somewhat slow, was considered the best for the purpose. Two of the four jaws were left set after the first piece had been machined so that in placing the

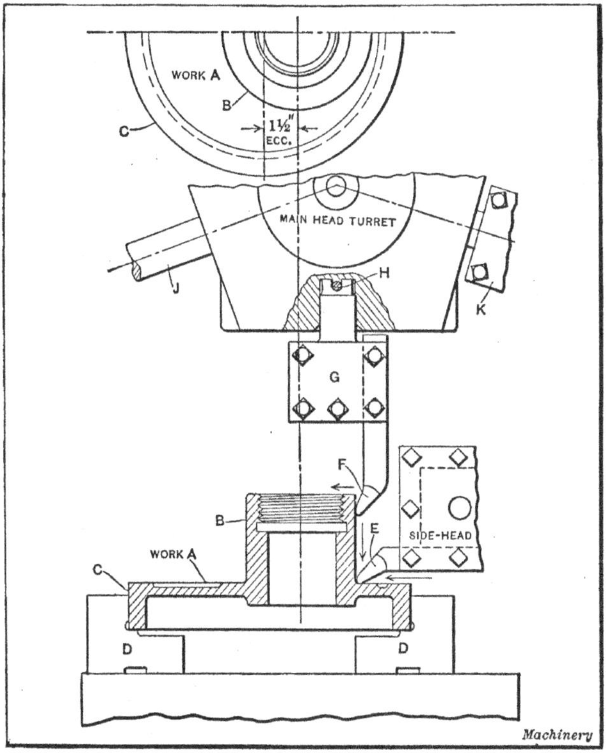

Fig. 6. Method of Machining Eccentric Hub on Vertical Turret Lathe

other pieces in position, these two jaws acted somewhat like a V, and greatly facilitated the setting up. A conical plug was used in one of the turret faces to assist in approximating the cored center of the hub. This plug is not shown in the illustration but its action will be readily apparent. The tool *F* is mounted in a tool-holder *G*, the stem of which enters the turret hole and is driven by the pin *H*. This tool is obviously used for rough-turning and facing the hub. While this operation is going on, the tool *E* in the side-head is used for facing. A boring-bar *J* is next used to bore the hole and while this is being done a finishing tool (not shown), in the side-head turret, is

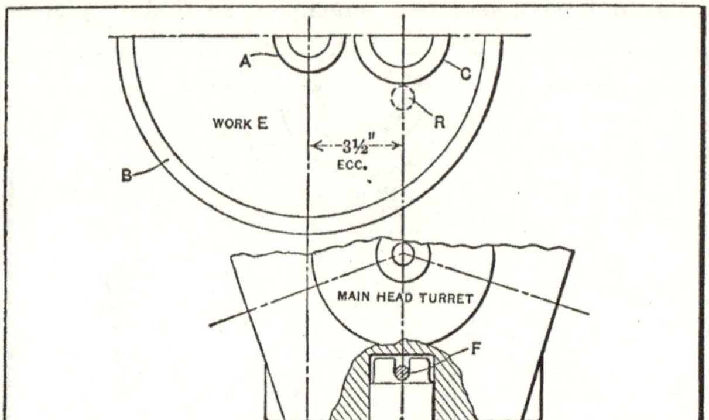

Fig. 7. Eccentric Turning Fixture for Vertical Turret Lathe

swung into place and is used for finish-facing and turning the hub. A tool-holder *K* in the main-head turret contains a grooving tool for the thread recess, while a thread chasing bar is used to cut the thread.

Eccentric Turning Fixture for the Vertical Turret Lathe. — The piece of work shown at *E* in Fig. 7 represents a condition somewhat out of the ordinary, in that the eccentric throw is $3\frac{1}{2}$ inches. By referring to the upper part of the illustration it will be noted that the small hub *A* and the rim *B* are concentric, while the hub *C* is eccentric, the amount previously stated. As there were a number of these pieces to be machined, and as the eccentric throw was so great, it was considered expedient to design a fixture on which they could be handled. Previous to the setting shown, the work had been machined on the periphery *B*, the edges of the flange had been faced, together with the hub, and the hole in the hub *A* had been bored and reamed. The body of the fixture *H* on which the piece is now held is of cast iron and is centered on the table by means of the plug *K*, while it is held in position by the T-bolts *L*. A hardened and ground steel plug *J* is located at the proper distance from the center of the table and on it the work locates. A plug *R*, shown in the upper view, strikes against the larger hub and acts both as a locater and driver for the work. The three clamps *N* are drawn down on the edge of the flange by the screws *M*, the lugs *O* on the fixture body being used to support the outer end. These clamps are of the type usually termed U-clamps so that they may readily be pulled back out of the way when releasing the work. A pocket *Q* under the large hub allows the end of the boring-bar to pass through the hole. An opening *P* allows easy removal of chips. The boring-bar *D* fits the turret hole and is driven by the pin F in the turret. This bar is made by the Bullard Machine Tool Co., and is provided with a set of slip cutters and reamers for any size hole within its capacity. (The construction of the bar was described in Chapter II.) The tool *G* in the side-head turret is used for facing the hub during the boring operation.

Fig. 8 shows a makeshift arrangement for machining the work *A* on a vertical turret lathe. In this instance the portion *B* is eccentric to *C* $\frac{1}{2}$ inch. The work is held by the inside in the jaws *D* of a three-jawed table, so that the portion *C* is concentric with the center of the table. A steel ring *E* is bolted eccentrically to the center of the table by the same amount as the desired eccentricity on the work. This ring is beveled on the inside to allow clearance for the turning tool *J*. This tool is held in a special holder *K* which has a shank running up into the turret hole and there driven by a pin *L* in the usual manner. A hardened and ground tool steel roll F is mounted on a shouldered screw which enters the rectangular holder *G*, a thrust washer *M* being interposed between the moving sur-

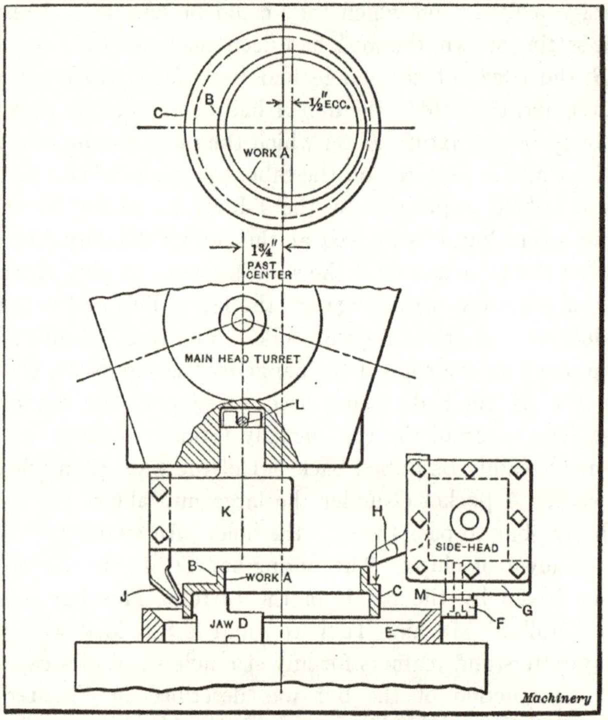

Fig. 8. Eccentric Turning by Means of Eccentric Guide Ring

faces. The two sides of the roll holder *G* are ground parallel, so that the roll will always be in a position parallel to the outside of the cam *E*. It will be seen that if this roll were not so located, there might be a possibility of cramp developing when pressure was applied. The tool *H* is, obviously, used for turning the eccentric. Attention is called to the fact that in the use of this device the operator's constant attention is necessary in order to produce accurate work. The cam roll F must be forced against the cam *E* by the pressure of the feed lever crank on the side-head, and the pressure applied must be sufficient to overcome any tendency of the pressure of the cut to push the roll away from the cam. It can, therefore, be readily seen that the operator's task is by no means easy in this instance.

Boring and Turning an Eccentric Ring Pot on the Vertical Turret Lathe. — The detail view to the left in Fig. 9 shows a plan of the ring pot *A*, the outer surface of which is eccentric

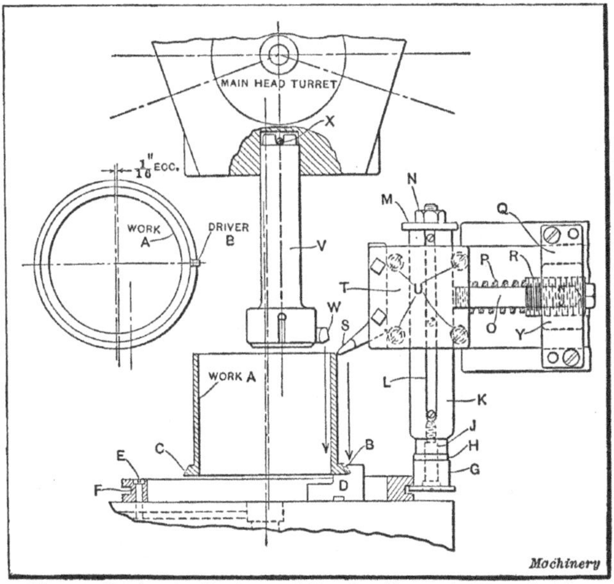

Fig. 9. Device for Boring and Turning Eccentric Ring Casting

to the hole by $\frac{1}{16}$ inch. Attention is called to the driver *B* which obviates the necessity of a great amount of pressure being used in setting up the jaws to hold the work, this lug being twisted around when placing the piece in position, so that it comes against the side of the jaw and greatly assists in driving the work. The jaws *D* are formed to the same bevel as the flange and the table is of the three-jawed type. The boring-bar *V* is of the same type as that previously described, being driven by the pin *X* in the turret. The tool *W* is, obviously, used for boring the concentric interior of the ring pot. A special cam ring having a groove F on its periphery is eccentrically set on the table and secured in position by the screws *E*. These screws pull up on shoes placed in the table T-slots. A special side-head *T* is used in connection with this device (the regular side-head turret being removed), and it is held securely in position by the four screws *U* which are tapped into it through the rear of the slide.

A hardened and ground tool steel bar *K* passes through this special block and is prevented from turning by the key *L*. A nice sliding fit is made between the bar and the block and the upper part of the bar is provided with a collar *M* and a nut *N*, in order to prevent its dropping out of position when the side-head is pulled back. The lower end of the bar is fitted with a shouldered screw *J* on which the flange roller *G* revolves. A thrust washer *H* is interposed between the roll and the end of the bar. A bracket *Q* having a shape somewhat in the form of a U is fastened to the side-head saddle, and bridges across the slide in the manner shown. On the under side of the bracket is the boss *Y* which is tapped out for the adjusting screw *R*. The pin *O* is screwed into the side-head and acts as a guide to keep the spring *P* from buckling. This spring is of $\frac{1}{4}$ inch square section and is very stiff. It will be noted that the pressure of the spring can be regulated to a nicety by means of the screw *R*. The tool *S* is used for turning the eccentric and it is held in place by the two screws shown.

In the operation of this device the side-head slide is moved forward by the pressure of the spring so that the roll *G* is kept in contact with the cam. The down feed of the side-head is then thrown into engagement and the tool *S*, following the movement of the cam, produces the eccentric on the outside of the ring pot. As the side-head moves downward the flange on the cam roll strikes the lower part of the groove F in the cam and allows the side-head to slip easily downward on the bar *K*. While the outside turning is in process the boring tool *W* in the main-head turret bores the interior of the pot concentric. Two tools *W* and *S*, acting in opposition to each other, tend to eliminate vibration in the shell and also assist in maintaining the sizes.

Eccentric Turning on the Vertical Boring Mill. — The work *A* shown in Fig. 10 is of cast iron and has a hole *B* which is 1.2 inch eccentric to the rim *C*. The machine on which this work is done is a vertical boring mill having two heads. The work is gripped by the inside of the rim in a four-jawed chuck,

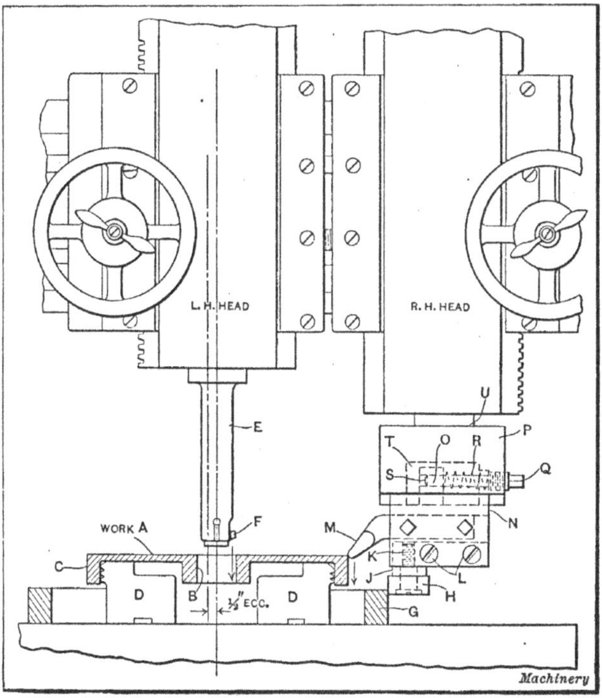

Fig. 10. Eccentric Turning on Vertical Boring Mill

two of the jaws *D* being shown in the illustration. These jaws are special and are set off center sufficiently to bring the hole *B* in the center of the table. A boring-bar *E* with a set of slip cutters F is used in the left-hand head to bore and ream the hole. This operation, however, does not take place while the outside turning is going on, as the boring speed would be too slow to be profitable. A tool-holder with a facing tool in it is used in place of the bar to face across the web while the eccentric turning is taking place. The tool-holder is then slipped out and the bar *E* placed in position, after which the table is speeded up and the hole bored.

A cam ring *G* is fastened to the table in the proper position to produce the correct eccentricity. A special device is mounted in the right-hand head, the body *P* being of steel and having a shank *U* which enters the hole in the ram. The lower portion of this block is dovetailed and to it the sliding tool-holder *N* is fitted. The lug *O* enters a recess in the block *P* and gives a surface against which the spring *R* thrusts. This spring is of heavy section and is adjusted by means of the screw *Q*, the end of which is turned down to fit the inside of the spring and prevent it from buckling. The end *S* pilots in a hole in the lug. The cam roll *H* is of tool steel hardened and ground and it revolves on the cylindrical portion of the shouldered screw *K*, the thrust washer *J* being interposed between the roll and the block. The holder is rectangular in shape and accurately fits a slot in the slide *N*. The two screws *L* hold it in place and incidentally tie the open side of the slide together. The tool *M* enters the upper portion of the slot and is held in place by the two screws shown. It can readily be seen that the action of the spring *R* tends to force the slide *N* forward, thereby keeping the roll *H* continually in contact with the cam *G*. As the tool *M* is held in the slide *N* it is obvious that it will follow the path generated by the roll and produce the desired eccentric.

CHAPTER XI

COUNTERBALANCED AND INDEXING FIXTURES

Conditions Governing Design of Indexing Fixtures. — Work of irregular form having several holes to be machined, and castings or forgings in which a single hole is to be bored at one side of the piece, require special fixtures upon which they can be clamped for the machining operations. There are a number of conditions which affect fixtures of this kind; for instance, the work may be small and of light weight, or it may be large and very heavy; it may have several parallel holes, or they may be at various angles, sometimes even converging to a common center; the part may be very much one-sided; it may have been partially machined in some previous setting, or perhaps is a rough casting or forging. All these things affect the design of the fixture, as does the type of machines to which the fixture is to be applied.

Counterbalanced fixtures are used only on the class of machines where the work itself revolves, such as engine lathes, turret lathes, boring mills and cylindrical grinding machines. Indexing fixtures, however, can be used on all the foregoing, and also on many other machines, such as drill presses, milling machines, planers, shapers, slotters, etc. Very frequently indexing fixtures are supplied with some provision for counterbalancing, when used on the first-mentioned group of machines. Sometimes this counterbalance is so arranged that it can be moved to suit different conditions, and, in other cases, it is made solid with the fixture and is carefully balanced at the time the fixture is made.

There are two things which must be carefully considered before any attempt is made to design a fixture, *viz.*, the machine most suited to the work in question, and the speed at which it will have to run in order to be thoroughly efficient. The latter is important if a counterbalance is to be used, because more care must be exercised in the balancing if the fixture is to be run at high speed. In addition to this, a high-speed fixture should be arranged with a suitable oil-guard, if cutting lubricant is to be used. It is evident, therefore, that these things should be thought of before the design of the fixture is started. This is one of the cases where "an ounce of foresight is worth a pound of hindsight," for when a device of this kind is finished and ready to put on the machine, it is very annoying to be obliged to hold up the work a couple of days while a guard is being made and fitted. Incidentally, it does not redound to the designer's credit, for it should have been thought of in advance.

Points on Fixture Design. — The following paragraphs call attention to the most important considerations in the design of indexing fixtures:

1. Rigidity is of prime importance in every kind of a fixture, and when the fixture is designed for use in a horizontal machine, the overhang from the end of the spindle should be made as short as possible in order to prevent vibration.

2. Positive location for the work and rapid-acting clamping devices of such a nature that it is not necessary to disturb them when indexing the fixture are of great importance. In this connection it might be mentioned that the author has seen indexing fixtures which were so designed that the clamps holding the work required loosening before indexing. Needless to say, this is very poor design.

3. A method of obtaining accurate indexing with provision for taking up wear at the locating points must be provided. Either taper pins or taper wedges are suggested.

4. Adequate protection from dirt and chips is necessary, so that errors in indexing will not be caused by imperfect contact at locating points. This point is of special importance on fixtures indexing in a horizontal plane, as, for example, those used on a milling machine or a vertical boring mill.

5. The method of counterbalancing is a feature of special consideration. If the counterbalance is to be a part of the fixture, the weight must be figured so that it will slightly exceed

the portion which it balances, in order that it may be drilled out or chipped off when assembled with the work in position, to obtain a smoothly running fixture. A sliding counterpoise is a very good arrangement when several pieces of a similar kind are to be machined, for it can be so proportioned that it will take care of varying conditions.

6. The rapidity of handling, both in clamping the work and in the operation of the indexing mechanism should be considered. As these two points make considerable difference in the cost of the fixture, they should be proportional (to a certain extent) to the number of pieces which are to be machined. In fact, the entire fixture should be designed with this point continually in mind, so that its cost as a whole will not be too great.

Counterbalanced Fixture for a Connecting-rod. — A very good example of a simple counterbalanced fixture is shown in Fig. 1. In this case the work *A* is a steel connecting-rod forging which has been faced on both sides and the wrist-pin hole in the small end has also been bored and reamed in previous operations. This fixture is used to bore and ream the large hole at a fixed distance from the wrist-pin hole and parallel thereto, the machine being a horizontal turret lathe. The body of the fixture *T* is of cast iron and is screwed to the spindle. The steel stud *M* is screwed into the body and is held in position by the pointed set-screw *N*. It is located at the correct distance from the center of the spindle and is turned down and threaded at one end to receive the U-washer *L* which is clamped down on the work by the hexagon nut shown.

The sliding V-block *D* is dovetailed to fit the slide *U* in the body of the fixture. This centers the work from the outside of the forging and it should be noted that it also has a holding down action due to the angularity of the V-block. This is clearly shown in the lower view. The lugs *H* and *J* act as supports for the swinging latch F, and the detent pin *K* keeps the latch in position. The block *D* is clamped against the work by the square-head set-screw *G*, located in the latch. The elliptical hole *E* is a finger hole by which the block is pulled back out of the way when removing and placing a new piece of work in position. The cored pocket *V* in the body of the fixture is provided so that the V-slide can be planed without difficulty, the pocket allowing a place for the tool to " run out."

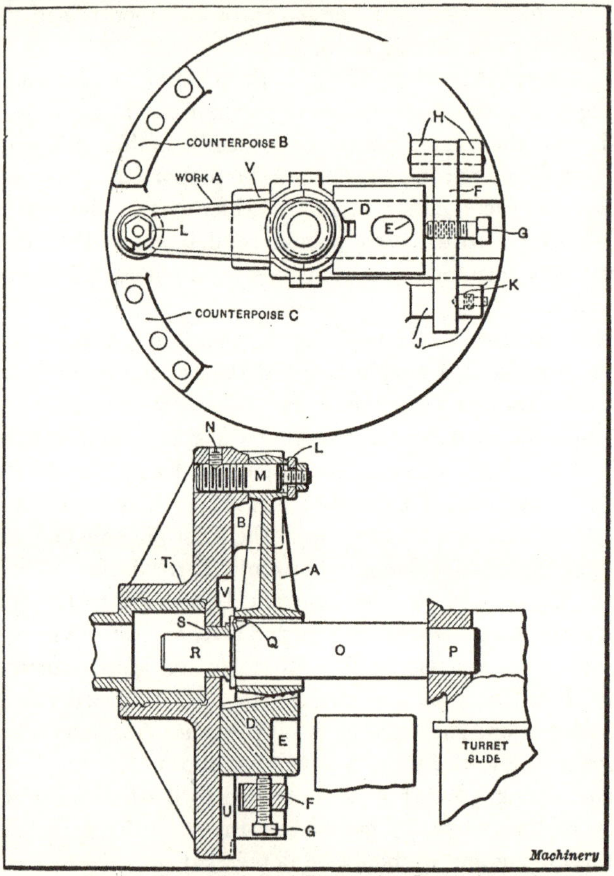

Fig. 1. Counterbalanced Fixture for Connecting-rod

The lugs *B* and *C* serve to counterbalance the weight of the latch and blocks on the opposite side and are at first proportioned so that they are somewhat heavier than necessary. After the work is put in place on the fixture it is carefully balanced by drilling into these lugs until sufficient stock has been removed to secure this result. Attention is called to the fact that the counterbalance is located as far from the center of the fixture as possible. This is done in order to make the weight more effective. A hardened and ground tool steel bushing *S* is located in the fixture body and acts as a guide for the boring-bar pilot *R*. The bar *O* is turned down at the rear end, as shown at *P*, so that it fits the turret hole. A recess in the fixture allows the boring tool *Q* to pass slightly beyond the work. A pilot bushing in the spindle might have been preferable, in that there would have been more space in front of the tool. If this had been done, however, the fixture would not have been self-contained.

Adjustable Counterbalanced Fixture for a Worm-gear Sector. — A condition which is somewhat out of the ordinary is illustrated in Fig. 2. The work shown at *A* is of two different sizes. The outside diameter is almost the same, but there is considerable difference in the weight of the two castings, one of them having a much wider rim than the other and also a heavier web section. Therefore it was necessary either to make the fixture in such a way that it would take care of both pieces, or else make two fixtures. As all other points in design would be the same, it was decided to make a movable counterpoise which could be adjusted to take care of the difference in weight.

The fixture body F is of cast iron, and is screwed onto the spindle as shown in the lower view. The locating stud *B* is shouldered and is held in place by a nut and washer on the inner end; the outer end is turned down and threaded to receive a hexagon nut which clamps the work through the U-washer *D*. The pointed set-screw *H* is set at a slight downward angle and forces the work over against the stud *G*. The partial section shows how the angular part of the stud tends to overcome any tendency to lift or strain the casting out of its correct position. The circular pad *E* forms a resting place for the work. The counterpoise *J* is a cast-iron segment resting on the two finished surfaces *O*. Two slots *K* permit the passage of the screws *L* which clamp the counterpoise in the desired place. An enlargement of the slots on the back of the body serves to prevent the hexagon nuts from turning when the weight is being adjusted. The lower part of the illustration shows the tool-holder *N* fastened to the face of the

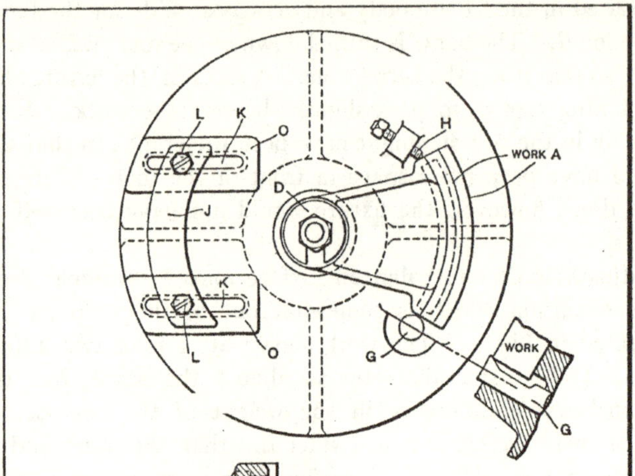

Fig. 2. Fixture with Adjustable Counterbalance

turret and ready to take a straight cut on the outside of the sector with the tool *M*. This arrangement was very satisfactory in every respect. The two sectors were put in position and the counterweight adjusted for each piece, the pads being marked by a scratch line so that the counterweight could easily be placed in either position.

Swinging Fixture for an Eccentric Piston Ring. — The work *A* shown in Fig. 3 is a cast-iron ring pot from which eccentric packing rings are turned and cut off. Previous to the machining operation indicated, these pots are rough-ground on the back of the clamping lugs *E*, and are also jig-drilled at these points.

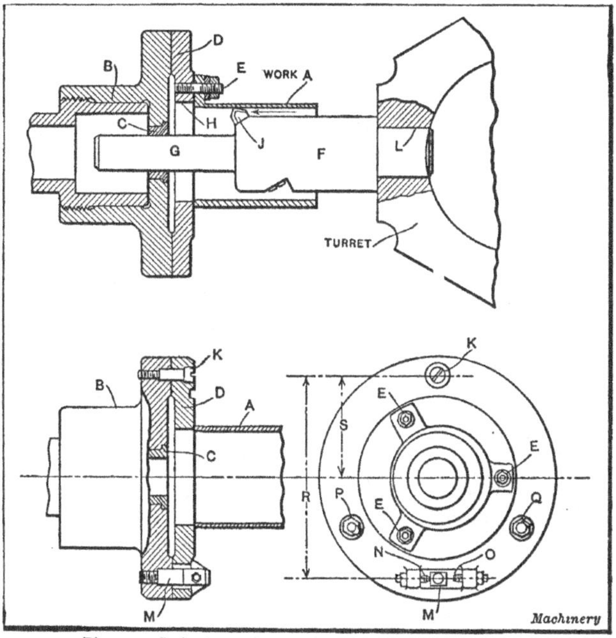

Fig. 3. Swinging Fixture for Eccentric Piston Ring

The customer for whom this fixture was designed did not wish to go to the expense of a special attachment for boring and turning eccentric at the same time. Therefore it was considered expedient to design the holding device so that it could be set over to produce the proper amount of eccentricity without re-setting the casting. The machine on which the work was done was a horizontal turret lathe of standard make.

The work *A* is located on the three studs *E*, and held fast by nuts and washers. The body of the fixture *B* is of cast iron and is screwed onto the end of the spindle. It is fitted with a bushing *C* which acts as a guide to the pilot end *G* of the boring-bar F. The swinging plate *D* is pivoted on the shouldered screw *K*, and it should be noted that this screw is set well down into the body of the fixture to give additional strength. An opening is provided in the plate *D* so that the square stud *M* in the body can extend out through it to act as a stop for the screws *N* and *O* in the swinging plate. These screws are provided with check-nuts so that they can be locked after being properly adjusted. In this particular case, the eccentricity of the ring was $\frac{1}{16}$ inch and, as the distance *R* on the fixture was made exactly double the distance *S*, only $\frac{1}{8}$ inch adjustment between screw *O* and stud *M* was required to obtain the correct setting. The studs and nuts *P* and *Q* are used for clamping the plate firmly to the body, the holes being slightly elongated to permit the necessary movement. In the operation of this device, the tool *J* in the boring-bar *F* was first used to bore the hole. The plate was then swung over against the stop and the turning tools machined the outside eccentric. Excellent results were obtained with this device.

Counterbalanced Indexing Fixture for a Carbureter Body. — The work *A* in Fig. 4 is a brass carbureter body, and the fixture for it is considerably out of the ordinary in regard to the counterbalancing. The shifting of the slide from one side to the other makes it necessary to arrange the weight in such a way that it can be changed quickly from one side to the other, as occasion may require. The machine used with this fixture is a Pratt & Whitney turret lathe. Prior to this operation the casting is milled on one side on a vertical milling machine. It should be noted that a lug is left on the casting at each end for clamping purposes; this is cut off after the machining operations have been completed.

The body of the fixture *R* is cast iron and it is screwed onto the spindle. The slide *S* is fitted to the dovetail *V* in the body of the fixture, and the indexing locations are obtained by a taper pin in the bushings *M* and *N*. The clamping of the slide in each position is effected by the shoes *O*; these are beveled to the same angle as the dovetail and are prevented from turning by the set-screws *Q* which bear against the flat portion of the shoes. The hollow set-screws *P* act against the ends of the shoes and hold them firmly against the dovetailed portion of

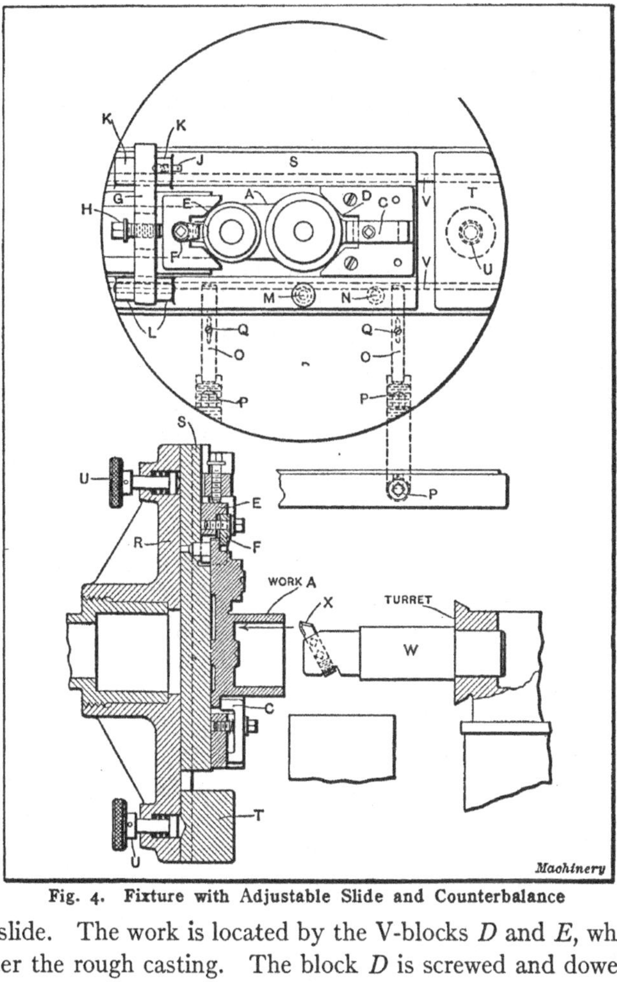

Fig. 4. Fixture with Adjustable Slide and Counterbalance

the slide. The work is located by the V-blocks *D* and *E*, which center the rough casting. The block *D* is screwed and doweled to the slide and contains the clamp *C* which bears against the lug on the casting. The block *E* at the other end is dovetailed on its lower side and fits the slide. The clamp F grips the lug and draws it up against the stud shown in the lower view. A latch *G* is pivoted on a pin in the lugs *L* and enters a slot in the opposite lugs *K* where it is retained by the detent pin *J*. The screw *H* forces the V-block against the casting. It will be noted that the latch is so hung that if the machine should be accidentally started while the latch is open, the latter would tend to close without damage to the fixture. This point in design is frequently overlooked and yet it is well worthy of attention.

The counterpoise *T* is of cast iron and is an easy sliding fit in the dovetail of the body, so that it can be removed without difficulty. It is countersunk on its lower side to receive the spring plungers *U*, which are located on opposite sides of the fixture. It is evident that this counterpoise must be transferred from side to side as the slide is indexed from one position to the other, so that the balance of the fixture can be preserved. The boring-bar *W* is shown in position in the turret; the tool *X* is used to rough out the pocket, as indicated by the arrow. This fixture was made up several years ago for an automobile factory in the eastern states, and it was used with good results for one season's production. It was abandoned only on account of a change in the carbureter design and not because of any fault in the fixture itself.

Counterbalanced Indexing Fixture for a Triple Cylinder. — The work *A* in Fig. 5 is a triple cylinder of cast iron with three bored and reamed holes equidistant about a common center *D*. The fixture was used on a horizontal turret lathe. The arrangement of swivel-plate, clamps and counterweight was made as symmetrical as possible so that very little change in the balance of the fixture would take place when indexing from one cylinder to another. The work was milled across the base and the holes *E* jig-drilled before the turret lathe operation took place. The body of the fixture is screwed onto the spindle and is made of cast iron. The swivel-plate which holds the work is also of cast iron and is pivoted on the hardened and ground steel stud *D* which enters the body. The three pins *E* act as locaters and drivers; they are in the plate and match the jig-drilled holes.

Three clamps F are used to hold the piece, and they are bent over on the ends so that they enter a slot in the plate, which prevents them from turning while the screws are being tightened. They are also slotted so that they can be pulled back out of the way when putting the work in position.

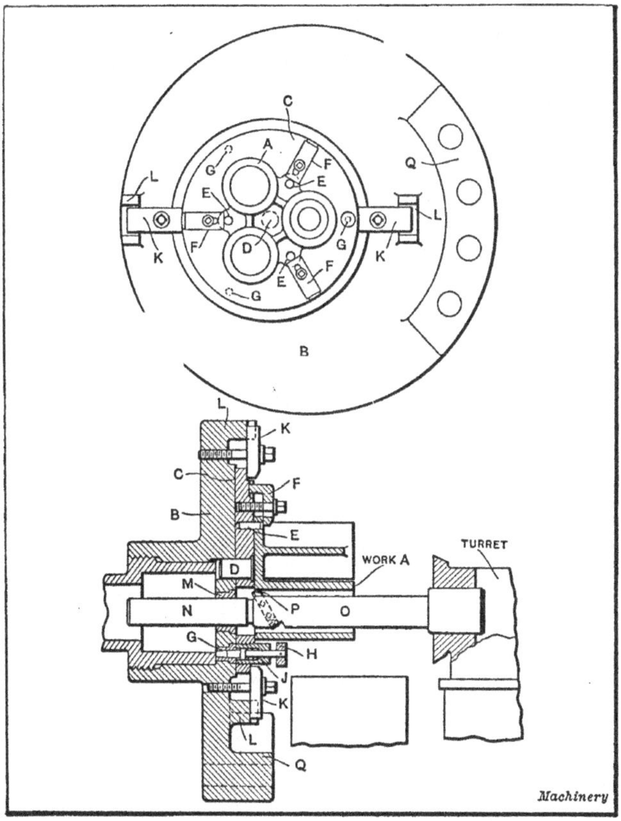

Fig. 5. Counterbalanced Indexing Fixture for Triple Cylinder

A hardened and ground steel plunger *H* (shown in detail in the lower part of the illustration) is tapered to fit the bushings *G* which control the indexed locations as the plate is swiveled. The steel bushing *J* acts as a bearing for the pin and also forms a pocket for the spring. The clamps *K* are loosened when indexing the plate but are set up again before the machining is done, so that rigidity and freedom from chatter are insured. The counterpoise *Q*, in this instance, is a part of the body casting. At first it is somewhat heavier than the parts to be balanced; it is then brought to the required weight by drilling out the stock after the work is clamped in position. Attention is called to the fact that the body *B* contains a bushing *M* in which the piloted end *N* of the boring-bar *O* is guided. It will also be seen that considerable chip room is provided for the tool *P* when it passes through the work. The results obtained by the use of this fixture were perfectly satisfactory.

Indexing Fixture for a Cast-iron Valve Body. — The work shown at *A* in Fig. 6 is a cast-iron valve body to be machined complete at one setting. It will be noted that the holes are at 90 degrees with each other, so that the fixture required indexing through an arc of 180 degrees during the process of machining. The casting is in the rough state when placed in this fixture. The cast-iron body *B* is screwed to the spindle, and its forward end is formed like the letter U to receive the swivel-bracket *C*, which holds the valve body. This swivel has a pair of V's which serve to locate the casting centrally in one direction. The swinging clamps F are pivoted at one end on the shouldered screws *H*, and are clamped against the casting by the action of the thumb-screws *G*. The work is centered in the other direction by the steel " bull-center " T, which is forced onto the shoulder of the bar *R*. This center is used in the rough cored hole before the clamps have been set up tightly, and while it is in position, all screws and clamps are tightened. The pointed set-screws *J* and *K* overcome any tendency of the casting to move during the machining operations. The bearings for the swivel are somewhat out of the ordinary as they are provided with a " take-up " for wear. The hardened and ground steel studs *D* are screwed into place in the swivel and are tapered on their outer ends to form the pivots. The threaded bushings *P* are milled hexagon on the ends and are tapered on the inside to fit the pivots. Headless set-screws prevent the bushings from turning.

The indexing locations are determined by the bushings *L*, which are forced into the swivel at the three points shown. These bushings are tapered on the inside to fit the pin *N*, which is also encased in the bushing *M*. A coil spring assists in keeping the pin in position. This fixture was not satisfactory due to the overhang of the spindle and a certain tendency to chatter caused partly by the excessive overhang and partly by imperfect balance when in either of the positions not shown in the

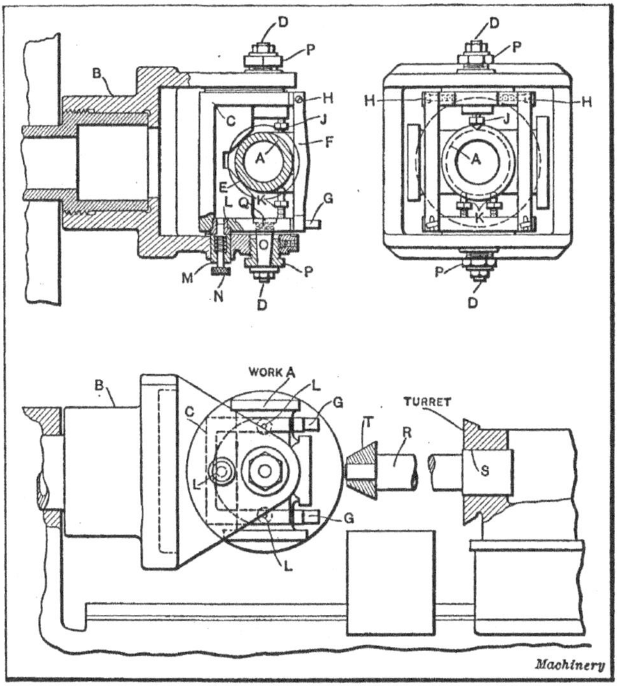

Fig. 6. Indexing Fixture for Cast-iron Valve Body

illustration. The author has, however, applied a fixture of very similar construction to a vertical turret lathe with perfectly satisfactory results, showing that the greatest fault was in the excessive overhang when used on the horizontal type of machine. Some of the details of construction in this fixture may, however, be applied to advantage on other work, and it is of considerable value in showing "what not to do," for we really learn more by failures than we do by successes, although a failure may be somewhat expensive.

Indexing Fixtures for the Vertical Turret Lathe. — Let us now take up the subject of fixtures which are designed for use on machines of the vertical spindle type. In the majority of cases, it is not necessary to provide counterbalances for this

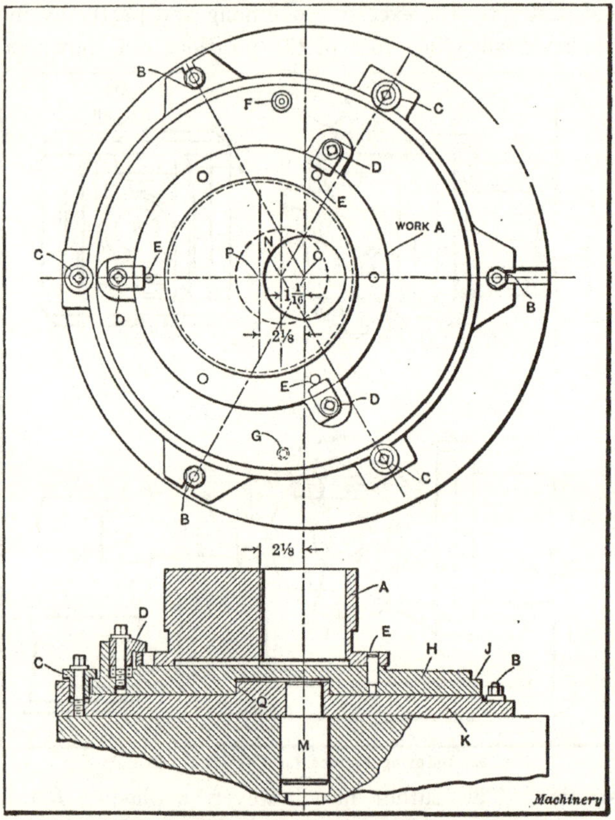

Fig. 7. Indexing Fixture for Turning and Boring Eccentric

type of fixtures because of the construction of the table bearing which always tends to center itself on account of its conical form. The great weight of the table is also of assistance in overcoming trouble which might be caused by heavy castings of an eccentric or one-sided form. It might be possible to have a manufacturing proposition come up which called for a great number of very heavy eccentric castings of the same kind. In a case of this kind, a counterweight might be used in order to save the machine from excessive strains and unequal wear on the spindle and table bearings. It is not, however, the usual practice to make any provision for counterbalancing.

Indexing Fixture for an Eccentric. — A rather peculiar fixture is shown in Fig. 7, this having been made for a vertical turret lathe. The work *A* is of cast iron and is an eccentric strap. In a previous operation this has been turned, faced, and recessed on the flange side, having been held by the solid portion of the body at this setting. The operation of drilling the flange holes in a jig also took place before the setting shown in the illustration. The work for which this fixture was designed is the boring of the eccentric hole, the facing of the top, and the turning of the outside portion.

The casting is located by the three pins *E* in the swivel-plate *H* and is clamped down by the three hook-bolts *D*. It will be noted that these hook-bolts are well backed up by bosses on the swivel-plate. The backing is cut away on one side so that the bolt can be swung around when placing a new casting in position. The coil springs underneath assist in releasing. This swivel-plate has the point *N* (shown in the upper view) for its center and swings on the boss *Q*. As point *N* is equidistant from *O* and *P*, it follows that when the swivel-plate is swung around 180 degrees, it will be in the correct position for turning the outside diameter. The three buttons *C* are set in the base *K* and act as clamps on the annular shoulders *J* of the swivel-plate. The base itself is held down to the table by the T-bolts *B*, which enter the table T-slots. It is centered on the table by the plug *M*. The indexing locations are determined by the bushings F and *G* in the base. These bushings are tapered to fit a taper spring-pin in the swivel-plate. It will be noted that this method of holding the work made it possible to perform all the turning operations in two settings.

Twin Cylinder Fixture for the Vertical Turret Lathe. — The casting shown at *A* in Fig. 8 is a twin cylinder which has previously been machined across the bottom, and in which the flange holes have been jig-drilled. The casting is placed on the sliding plate *F* and is located by the two pins *D* which are diagonally opposite each other. The four clamps *E* are used on the flange to hold the work in position. It will be noticed

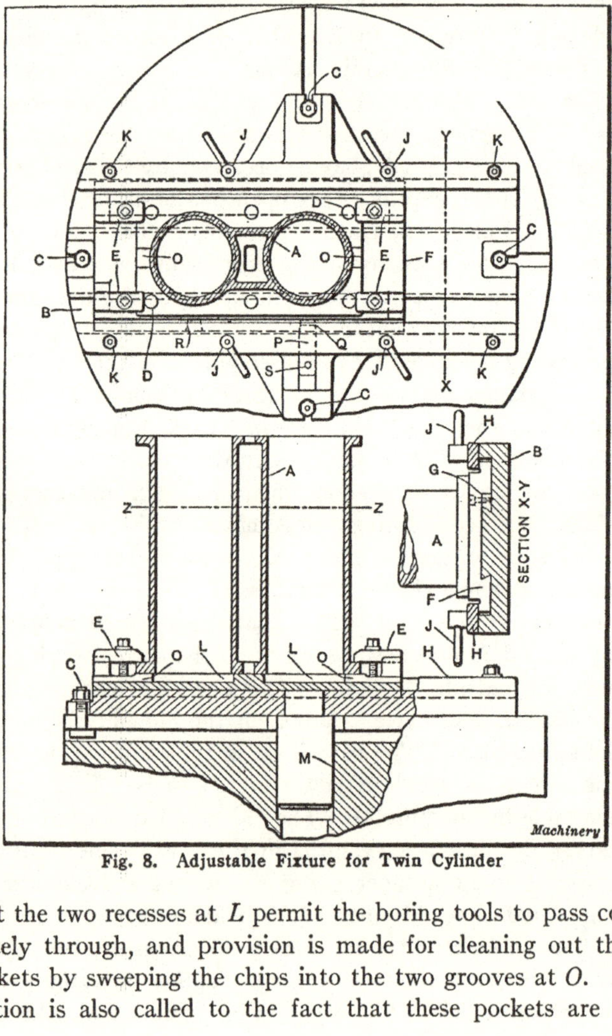

Fig. 8. Adjustable Fixture for Twin Cylinder

that the two recesses at *L* permit the boring tools to pass completely through, and provision is made for cleaning out these pockets by sweeping the chips into the two grooves at *O*. Attention is also called to the fact that these pockets are not bored entirely through the slide, for if this were done trouble might be experienced with chips or dirt. The slide is dovetailed on its lower side, as shown in the section *X–Y*, and provision is made for take-up by the gib *G*. The base of the fixture *B* is bolted to the table by the T-bolts *C* and is centered by the stud *M*. The steel strips *H* are so fitted that a sliding fit is insured when the binders *J* are loosened. The indexing is governed by the taper wedge shown at *P* in the upper view. This wedge enters the two slots *Q* and *R* which are tapered to receive it. The pin *S* is used to pull the wedge out of engagement. It will be noted that the view of the work shown in the upper illustration is a section taken on the line *Z–Z*, in order to make the construction more apparent.

CHAPTER XII

INFLUENCE OF CHIPS ON THE DESIGN OF TOOLS AND FIXTURES

Effect of Chips. — In the design of all kinds of cutting tools, jigs, fixtures and even machine tools, the chips produced by the cutting action have an influence which is of great importance. In these days when the amount of metal removed in a given time is so great in comparison with what it used to be, the importance of the matter is especially emphasized. In the design of machine tools of all kinds there are a number of points which may be traced back directly to the disposal of the chips, for example, the protection of running or sliding surfaces, such as the ways of a turret lathe. Some type of felt "wipers" are generally used for this purpose, as it is out of the question to arrange a permanent guard. When machines are used on cast-iron work some protection is greatly needed, as cast iron contains more or less sand which has an abrasive action and therefore causes excessive wear on moving parts. Cast-iron dust containing sand is nearly as destructive as emery. The author has seen the ways of a turret lathe so badly worn when used in cast-iron chucking, that the depression could be easily seen and even felt with the fingers in some places. Felt wipers on the front end of the turret slide would have prevented a great deal of this trouble.

Gear guards are sometimes found necessary on account of flying chips, even when the gears are in a position which would not require guarding for the sake of the operator. Pans are now cast as a part of the machine bed in many cases, and adjustable guards are provided in addition, to catch chips and direct them into the pan. In the design of chucks and collets the influence of the chips is again apparent, for the chuck jaws are made of such lengths that they keep the scroll well covered under normal conditions, while collets are protected by their general construction. In milling machine construction the slide is well protected by its general design, as there are no exposed ways on which chips may find lodgment. The table, however, with its T-slots receives the chips, and provision for cleaning is therefore necessary. The slots, being open at both ends, may be easily brushed out. Table gears are in such a position that no trouble is possible here. Index heads for the milling machine have all their working parts well protected so that there is no danger of chip troubles on these mechanisms.

Chips and the Design of Cutting Tools. — As cutting tools are the great producers of the chips in question, it is apparent that the matter of taking care of the chips must be carefully considered in their design. The flutes in a drill, for instance, are not made for ornamental purposes, but are a necessity in order to provide a path by which the chips may find their way out of the hole. Cutting angles of various tools, their clearance and their general shape all have to do with the making and disposal of chips. On steel work from the bar, when a roller turner is used, the angle of the cutting tool controls the form of the chip to a certain extent, and on heavy reductions of stock when a continuous spiral chip is formed, more or less clogging of the turning tool sometimes takes place. In order to prevent this trouble the Jones & Lamson Machine Co. brought out their chip-breaking turner a few years ago. Other types of tools are affected by the chips in one way or another, those working on internal surfaces being somewhat more susceptible than those used externally, on account of the difficulty of getting rid of the chips after they are made. Speaking generally, internal tools, such as boring-bars, which are to be used on steel work, require much more clearance for chips than those which are to be used on cast iron or other material having a short chip. Many a tool has proved a failure on account of insufficient chip clearance.

Chips and Jig and Fixture Design. — There are several points in the design of jigs and fixtures for interchangeable work, which are of interest in their relation to the chips; for example, locating points or surfaces should be so arranged that the presence of chips will not cause variations in the location of the work, and they should be easily accessible for cleaning; threaded clamp screws and other movable parts should be protected, as far as possible, in order to prevent trouble in operating; indexing fixtures should have the indexing parts so designed that there can be no chips at important points; openings should be made in the jig or fixture so that the accumulation of chips incident to the cutting operations can be quickly removed.

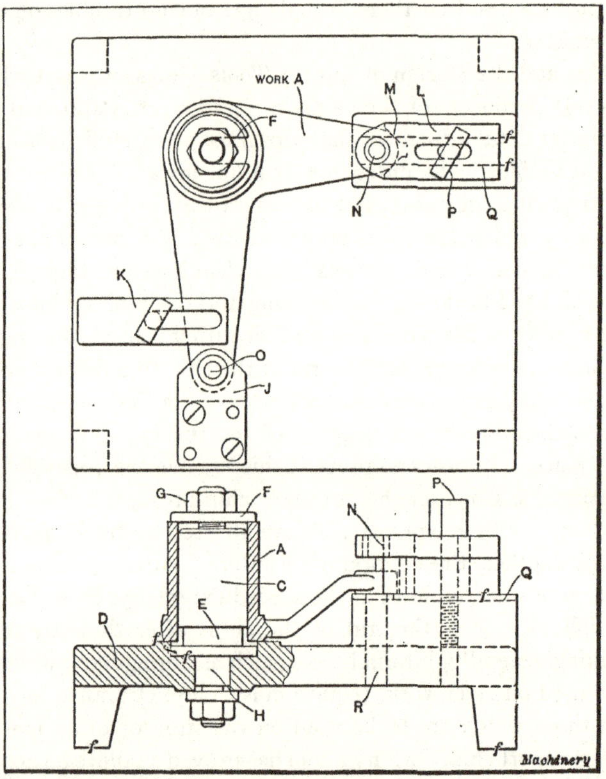

Fig. 1. Example of Drill Jig Design where Provisions for Chips are only partly considered

Fig. 1 shows two views of a drill jig for the bell-crank *A*. Prior to the operation performed in this jig, the piece has been chucked and the large hub turned, and the bosses have been faced. This example shows a peculiar inconsistency, in that the chips have been provided for in some portions and neglected in others. The casting is located on the stud *C* which fits the previously chucked hole in the work. This stud is shouldered at *H* and kept in place by the nut and washer shown. The U-washer *F*, in connection with the nut *G*, makes rapid removal easy. The shouldered portion of the locating stud sets slightly below the finished boss on which the hub of the casting rests, and is relieved at *E* in order to prevent chip troubles. As slight variations in center distances are permissible, the drilled hole at *N* is centered on the boss at the end of the bell-crank by mounting the bushing plate *M* on the sliding V-block *L* which is tongued to fit the slot *Q*, the thumb-screw *P* being used to lock the block in position. The other bushing *O* is fixed in plate *J* and a slotted clamp *K* holds the work down.

One of the peculiarities of this jig is the fact that although hole *R* has been made slightly oversize for the egress of chips formed in drilling the boss, no thought has been given to the trouble which might be caused by chips packing into the slot in the sliding block or the groove in which it travels. As a matter of fact the chips packed into the slot in such a way as to cause a great deal of annoyance, until the operator, who seemed to be of a more practical turn of mind than the tool designer, cut out a piece of tin to cover the slot; this helped the matter considerably. There was, however, no remedy for the packing in of small particles of metal under the block itself, except the removal of the sliding block from time to time to clean out thoroughly the groove in which it moved. Failure to take proper care of this matter caused angularity in the holes, due to the "cocking" of the block. This trouble was all caused by chips and could have been avoided by using a dovetailed block with a backing-up screw or something of a similar nature. It will be seen that in this particular case the tool designer's forethought took into consideration the location of the piece, but overlooked the production of chips in their relation to moving parts.

Provision for Chips in an Index Milling Fixture. — In milling machine work the chips produced are inclined to be rather fine and are frequently very abundant, so that precautionary measures are of the greatest importance. A point which should not be overlooked is that a milling machine fixture is practically a part of the machine itself when it is in use, and therefore must be readily cleaned without removal from the table. Oftentimes a fixture is completely buried in chips and a designer should bear

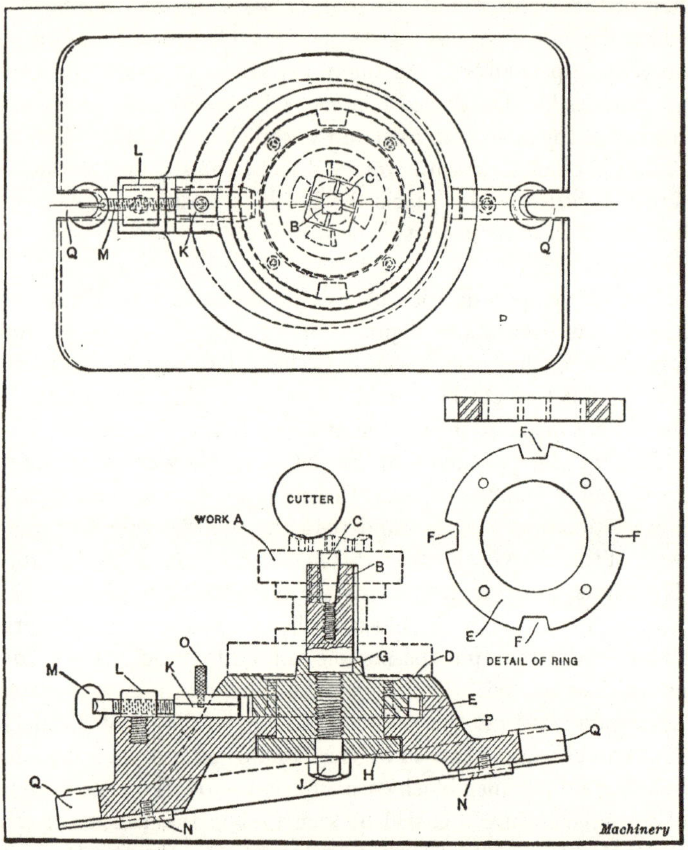

Fig. 2. Indexing Milling Fixture well guarded from Chips

this continually in mind, so that proper precautions will be taken to avoid pockets, recesses, etc., which might be difficult to clean. In indexing fixtures special care must be taken so that there will be no inequalities of the indexing due to the presence of chips.

Fig. 2 shows an index milling fixture for an automobile clutch gear *A*, in which a square hole has been previously broached. The work done in this operation is the milling of the angles of the clutch teeth. The work is located on the square stud *B* which is split in four places on its upper end and expanded by the tapered screw *C* in order to fill the hole and avoid any possibility of chatter which might be caused by a loosely fitting stud. The split feature also acts as a clamp. The stud is shouldered at *G* and screwed into the upper part of the indexing portion *D* on which the face of the larger gear rests. The index ring is screwed to the under side of this piece and the stem of part *D* extends down, having a bearing in base *P*, where it is secured by the collar and bolt *H* and *J*. The base is provided with keys *N* which fit the slots in the table, and slots *Q* for the holding-down bolts. The index ring (a detail of which is shown at the right of the illustration) is of steel and is provided with four angular slots *F*. The indexing bolt *K* is correspondingly tapered, and is tongued at the sides to fit the base. A pin *O* provides the means of withdrawal; it will be noted that the necessary movement is not sufficient to expose the angular point and thus allow chips to enter. Thorough protection is afforded both the index ring and bolt by the overhanging portion of the casting *D*. The bolt is forced into position by screw *M* acting in the swivel *L*. This fixture is an exceptionally good one, requiring very little care and having ample protection against chips.

Another Example of a Milling Fixture. — The work shown at *A* in Fig. 3 is a steel hub which has been previously bored, reamed, turned and faced. The requirements for the milling operation are that the four slots must be milled at right angles to each other and perpendicular to the axis of the hub.

The base *H* of the fixture on which the work is held is provided with two keys F which fit the table T-slots in the usual manner, while slots *J* are used for the hold-down bolts. Stud *B*, on which the work is located, is shouldered at *G* and forced into the base. Chip troubles are here prevented by allowing a considerable amount of clearance between the lower end of the hub and the boss in which the stud is held. The fixture is also of very open construction, making cleaning easy, as all surfaces can be readily reached with a brush. The two uprights are a part of the base casting and are finished at *C* to receive the flange of the casting. Clamps *D* are set up by means of screws *E*. It will be readily seen that there is no chance for chip troubles on these

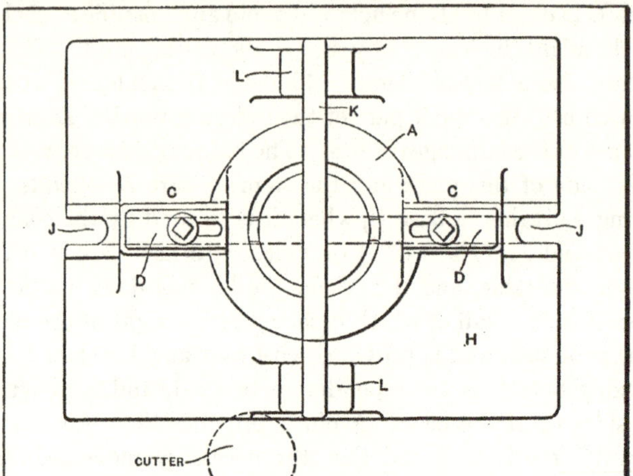

Fig. 3. Another Satisfactory Milling Fixture

surfaces on account of their accessibility. The indexing feature of this fixture is somewhat unusual, as no part of the fixture itself changes its position. After the cutter has cut the first two of the slots in the end of the hub, the clamps are loosened and the work is revolved on the stud until parallel *K*, guided by lugs *L*, can be pushed into position. The clamps are then set up lightly and the parallel removed, after which the clamps are tightened. This fixture is very satisfactory in every respect and is much less costly than an indexing fixture having a revolvable portion would have been.

Avoiding Chip Troubles in Chucking Work. — In the machining of work in chucking machines, turret lathes and the like, there are fewer chances for chip troubles in the holding devices, because these are in motion, and this has a tendency to throw the chips out of the way by the action of centrifugal force. Nevertheless, fine particles of sand from cast iron or steel and fine pieces of chips may cause considerable trouble unless proper precautions are taken.

The upper illustration in Fig. 4 shows an automobile piston *A* which has been previously machined on the outside. In this setting it is to be held by the outside surface and bored at *B*. It is held in the soft jaws *C* of a three-jawed chuck bored to receive it, a recess being cut at *D* so that locating troubles will be obviated. There is nothing unusual about this procedure, as it is common practice. A point in design which is frequently overlooked and which sometimes causes serious trouble is the tail *E* of the jaw. Unless this is made long enough to cover the operating scroll of the chuck perfectly, chips and dirt may find their way into the mechanism and do considerable damage. The author has seen a chuck of this type with improperly designed jaws so badly clogged up with chips that it was impossible to use it until it had been taken apart and cleaned.

The work *F* shown in the middle illustration is a bevel gear which is to be machined on the angular faces. The split bushing *G* is forced back on the taper arbor *H* by the nut *J*. The arbor itself is held in a special nose-piece on the spindle and has a shoulder *L* by which the longitudinal location of the work is insured. The relieved portion *K* of the arbor is of assistance in grinding the taper and also a precaution against dirt and chips. If used on cast-iron work, the split bushing should be taken off and thoroughly cleaned occasionally, as chips will find their way

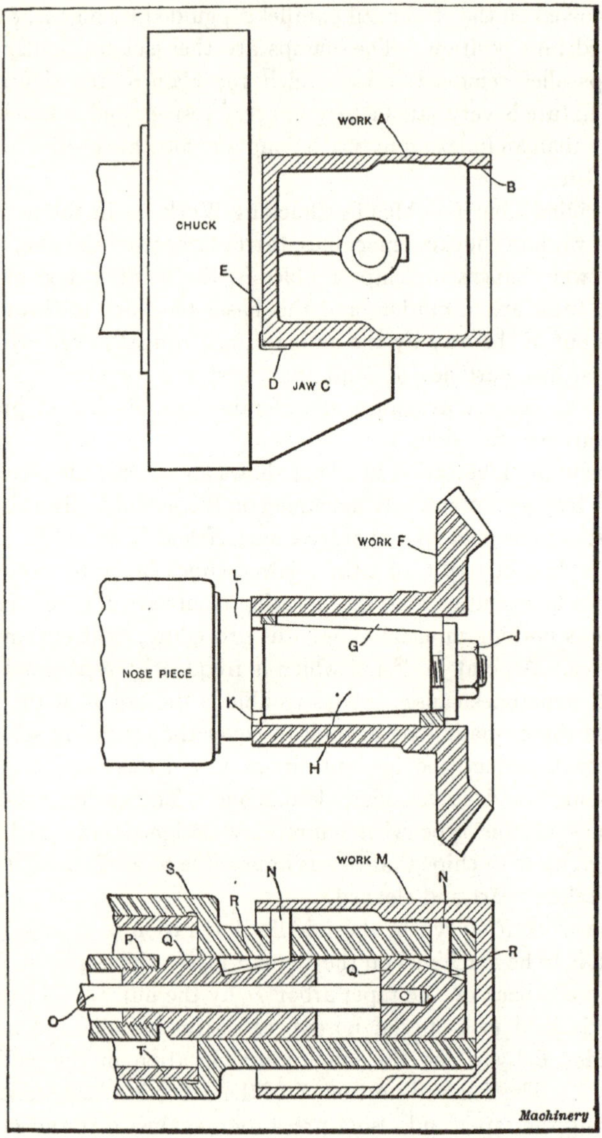

Fig. 4. Methods of Holding Work in Horizontal Chucking Machines, showing Means for Guarding against Chips

into the slots and between the bushing and the arbor, thus causing eccentricity. Troubles with expanding arbor inaccuracies may frequently be traced to this source.

The expanding-pin chuck shown in the lower part of Fig. 4 was designed for holding the piston *M* by the cored interior. Six radial pins *N* were dovetailed at their inner ends and were controlled by the dovetail-slots *R* in the cams *Q* which were operated by the rod *O* and the sleeve *P*, respectively, from the rear end of the spindle. The body of the chuck *S* was screwed to the spindle as shown. This chuck is an excellent example of the troubles which may be caused by cast-iron dust, for although mechanically correct, its construction was such that particles of sand, fine chips and other foreign substances became wedged in the dovetails in such a way that its operation was seriously impaired. A leather washer in the open end helped matters considerably by preventing the dirt from getting into the mechanism. In several instances before this was done, the dovetail portion of the pins was broken off by the operator on account of the excessive power needed to operate the mechanism.

Avoiding Chip Troubles in Vertical Turret Lathe Work. — In vertical turret lathe work much greater care must be used in the design of tools and fixtures than in horizontal work, as the chips naturally fall down onto the table or fixture. The table chuck on this type of machine is arranged with a set of permanent sub-jaws, lying flush with the table top and provided with a series of slots in which the upper jaws are fastened. T-slots in the permanent jaws permit quick and easy radial adjustment to approximate diameters. The scroll is well protected from chips, as the permanent jaws extend out to the edge of the table, there being a scroll controlled movement of about two inches which is sufficient to take care of any required condition.

When regular equipment and standard jaws are used there is no necessity for special precautions regarding chips, but when special tools and fixtures are used, considerable care must be taken in the design so that the devices can be easily cleaned. Movable clamps of all kinds must be so arranged that the falling chips will not clog and interfere with their use. Fig. 5 shows a

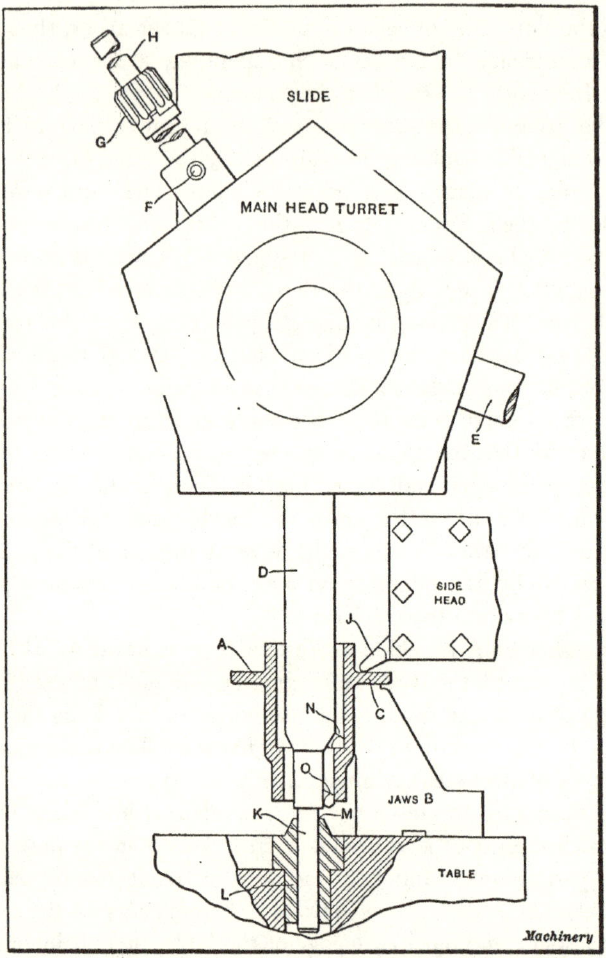

Fig. 5. Well-designed Device for Vertical Turret Lathe Work

hub at *A* which is to be bored, faced and turned on the upper end. The work is held in a set of special jaws *B* by the outside of the hub, the flange resting on the three points *C* on the upper parts of the jaws. In order to increase production it was desired to use multi-cutting bars as shown at *D*, and it was considered advisable to pilot the lower end of the bar in order to make it more rigid.

A bushing *L* was fitted to the center hole of the table and was used as a guide for the pilot *K* of the boring-bar.

As any chips falling down and wedging around the bar would do a great deal of damage, the end *M* of the bushing was made conical, while the hole instead of being rounded or chamfered as is usually the case, was ground square and sharp. The result of this procedure was that the greater number of chips did not remain on top of the bushing, but fell to the table where they did no damage. The sharp corner in the hole did not permit any

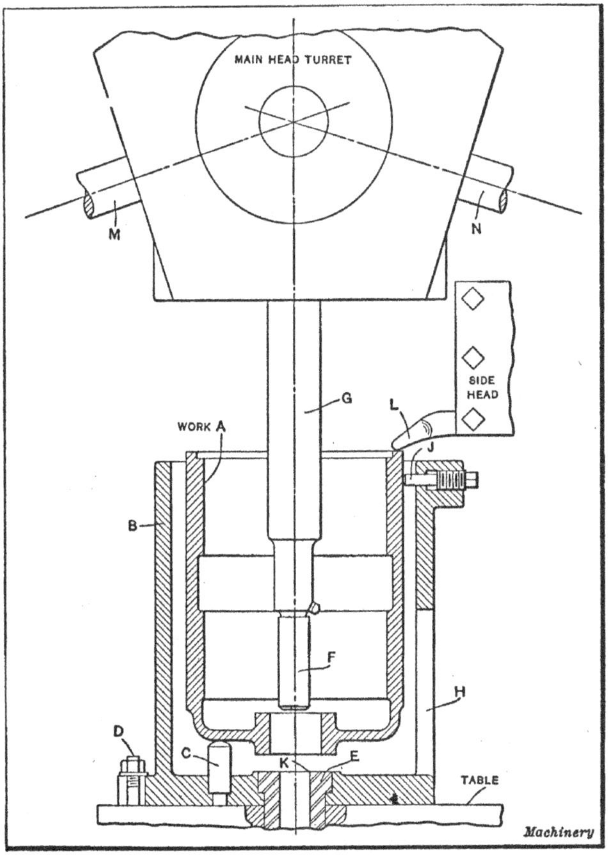

Fig. 6. Another Device for Vertical Turret Lathe where Due Regard has been given to Removal of Chips

wedging action or drawing in of chips around the bar at this point. Another bar *E* used for finishing was piloted in the same manner, while the floating reamer *G* was hung loosely at *F* and piloted by the stem *H*. The side-head tools were used in connection with the boring- and reaming-bars as indicated by the tool *J*.

In vertical turret lathe work heavy castings are frequently machined which sometimes require the use of a pot of some sort as a holding device. In cases of this kind it is very important to provide means of cleaning, both because of the rapid accumulation of chips in a pot of this kind, and because of the necessity of keeping all locating points clean.

Fig. 6 is a good example of a piece of work *A* held in a pot casting *B*. This casting is centered on the table by means of the bushing *E* which fits the center hole, the fixture being held down by the bolts *D* in the table T-slots. The work rests on three points, these being pins *C* in the base of the fixture. Attention is called to the fact that the tops of these pins are rounded so that chips will not lodge there. A series of set-screws at top and bottom are arranged as shown at *J*. These set-screws are placed 120 degrees apart. It will be noted that the thread of the screws is entirely protected from chips and dirt, so that no trouble is experienced in operating, as might be the case if a regular set-screw were used with an unprotected thread. Two boring-bars and a reamer are piloted as in the preceding instance, these tools being indicated at *G*, *N* and *M*. The edge of hole *K* in the bushing is made sharp to receive the pilot F and the top of the bushing lies flat and flush with the fixture. There are three cored openings *H* in the walls of the pot, these being large enough to admit the hand of the operator, so that he can readily clean the locating points *C* and the bushing *E*, and free the entire interior of the fixture from chips without trouble.

The examples illustrated in this chapter have been chosen for their simplicity, in order to point out the importance of the subject even under the simplest conditions. In fixtures of a more complicated nature, the question of chips becomes even more important, but the same principles can be applied.

CHAPTER XIII

PROVIDING FOR UPKEEP IN THE DESIGN OF CUTTING TOOLS

Upkeep of Cutting Tools. — It is of primary importance in the designing of all kinds of cutting tools to bear in mind the fact that all cutting edges must be occasionally sharpened by grinding, and that the re-grinding of the tools brings into use portions which have previously been inactive. This matter deserves the most careful consideration, in order that the design may be such as to give the longest life to the tools consistent with reasonable first cost. In the majority of cases the conditions are such that a little forethought will result in a material improvement in the "upkeep." There are occasional instances, however, when it is very difficult to construct tools so that they will not be seriously affected by re-grinding, as for instance when certain kinds of forming tools are required. A final finishing tool taking a light scraping cut may sometimes be found necessary in cases of this kind, and as this tool will seldom require re-grinding because it has so little to do, the work can be held very close to size.

Points in General Design of Cutting Tools. — The designer of cutting tools should have a wide experience and a broad knowledge of the actual conditions under which tools are to be used. He must be capable of thoroughly analyzing the situation and determining the type of tool best suited to the work in question. Examples which illustrate the various types of cutting tools in common use will be described in the following, and attention will be called to features in which there is a chance for improvement, with suggestions tending toward increased life in the tools.

1. The number of pieces to be machined is one of the first things to be considered, as it has a decided influence on the type of tool to be used. A tool designed for machining a small number of parts should be made as cheaply as possible, and the upkeep

is not of great importance; but if a great many pieces are to be handled the first cost is not as important as rapidity of action and provision for upkeep.

2. The various materials to be cut — steel, cast iron, brass, etc. — require tools of different shapes, and the clearance angles must be such as to turn the chips to the best advantage.

3. Sufficient clearance for chips is essential in all cases, but should receive special attention when tools are to be used on steel. Internal cutting tools are naturally more subject to chip troubles than those used externally.

4. A copious supply of cutting lubricant is of great assistance in prolonging the life of the tool, as it carries away the heat generated by the cut. If surfaces which are to be machined are so situated that they cannot readily be reached by regular methods of lubrication, it may be found advisable to provide oil channels in the tools themselves.

5. Particular attention should be given to portions of tools subject to slight changes in contour due to re-grinding. Forming tools, shoulder tools, taper reamers, etc., are especially subject to these variations when re-ground. Methods of overcoming these difficulties will be cited in this chapter.

6. The use of standard sections is advisable when tools of rectangular stock are required, as machining is thereby obviated. It is good practice to use sections such as are carried in stock in the factory where the tools are to be used, so that replacements can be quickly made.

7. It should be remembered that "sledge-hammer" methods of adjustment are not conducive to accurate work. Backing-up screws should be provided wherever possible, so that fine adjustments can be made by means of the screws.

8. Generous sizes of stock should be used for cutters whenever possible, as heavier cuts can be taken with tools of large size and they radiate the heat more rapidly, so that the production is increased and the upkeep facilitated. It is obvious that finishing tools may be made much lighter than those used for roughing, as the cuts taken are correspondingly less.

9. Rigidity is an important point and must be carefully con-

sidered in regard to the tool itself, its overhang and the method of holding it. Chatter is fatal to any kind of cutting tool.

10. The original cost of the tool is influenced by many factors. There are many instances when it is advisable to make two or more tools of the same kind, so that replacements may be quickly made in case of accident and so that the cost will be less for each tool than if made at different times.

Forged Tools and Similar Types. — Although forged tools are the simplest of the many varieties of cutting tools, they are greatly abused at times in their application, and much unnecessary expense is incurred by designing them in such a way that re-forging is the only salvation after the tool has been ground a few times. Take, for instance, the forged tool *A* in Fig. 1, which has an offset *B* of $\frac{5}{8}$ inch and a length *C* of 6 inches. Tools of this general type are frequently used on both horizontal and vertical turret lathes. They are very useful at times on account of the facility with which they can be used to cut in either direction. The operation of re-forging tools of this type is simple and comparatively inexpensive. A substitute for this tool may be made as shown at *D*, this type being straight and set angularly in a special tool-block and secured by two screws *E*. The upkeep is thereby made much simpler, as straight stock is used. It may be argued that the cost of the special tool-block will more than offset the extra forging required on tool *A*, and this is true to a certain extent, but when multiple cutting tools are used, a special tool-block is necessary in any case, and it is no more trouble to make it this way than straight. For horizontal turret lathe work, facing and turning tools of this type are much used.

Tool F is another type frequently used for cutting an internal recess or groove as shown at *G*. The size is soon lost by re-grinding so that re-forging becomes necessary. In equipments designed for machining a large number of pieces an efficient substitute may be provided as shown by the bar *H*. The end of this bar is cut out in the form of a five-degree dovetail and a tool *J* made to fit it. The bar is split as far back as the drilled hole *K*, and screw *L* is used to draw it together, thus securely binding the tool in any desired position.

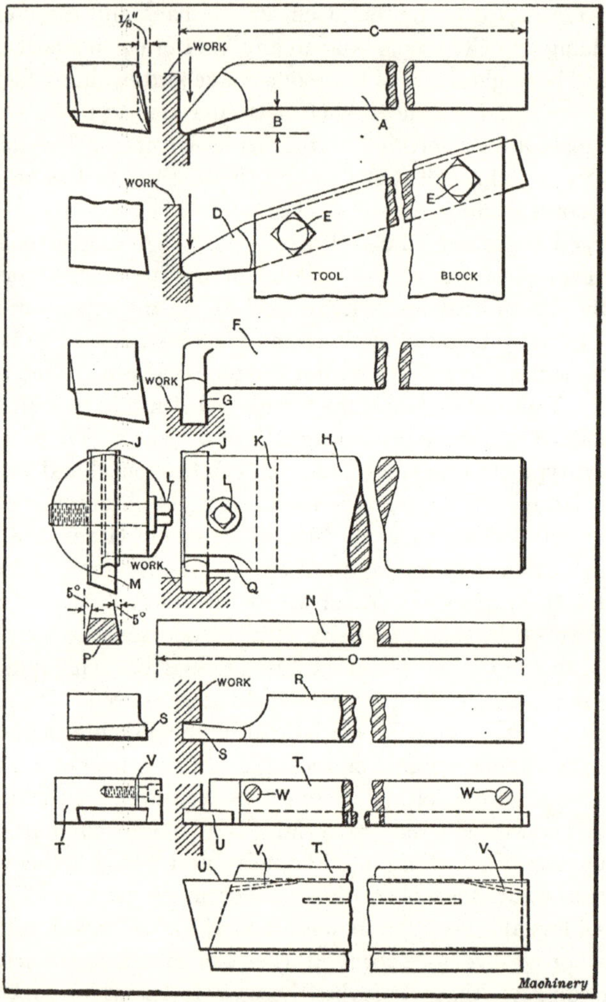

Fig. 1. Forged and Inserted-blade Turning and Cutting-off Tools

Attention is called to the manner in which the bar is cut away at *Q* for chip clearance. Tools may be planed up in a strip, as shown at *N*, with a length *O* from twelve to twenty inches and afterward cut off in lengths to suit. The section *P* shows the shape of the strip before grinding. Very little grinding is necessary to produce the required shape. A decided advantage in the use of a bar and tools of this type is the ease with which cutters of various sizes may be made, and the cheapness of upkeep.

One of the most common types of tools is shown at *R*; this is generally spoken of as a parting or cutting-off tool. The life of a tool of this kind is longer than some of the others mentioned when used on small bar work or thin castings, because the cutting portion *S* is quite long and will stand a good deal of grinding before it becomes necessary to send the tool to the blacksmith shop. Another cutting-off tool is shown at *U* and consists of a strip of high-speed steel, ground to fit a special holder *T*. This holder is split at each end at *V*, and two screws *W* hold the tool in place. Considerable economy results from an arrangement of this kind, and the holder may be bought in various sizes. It is one of the products of the Pratt & Whitney Co., Hartford, Conn., and is protected by patents.

External Grooving Tools. — External grooving tools may be required for wide or narrow grooving and may be spaced close together or several inches apart. They may be used in groups of from two to eight or ten, and may be all alike with equal spacing, or entirely different from each other. The author has selected two designs to represent this type of tooling, both of these being shown in Fig. 2. The one shown at *A* is designed to make two wide grooves on the exterior of a large cast-iron drum as indicated in the illustration. The tool-holder is of tool steel and has a shoulder at *G* which can be brought up against the tool-block, thus acting as a support and giving additional stiffness. Two holes are drilled at *B* and slots *C* are cut to permit a clamping action through screw F. Attention is called to the fact that the slots are not cut entirely through the holder, so that the lower part remains tied together. The tools *D* are of high-speed steel and are set in the holder at an angle of 15 degrees to give the necessary clearance. The screws *E* permit of easy adjustment. An arrangement of this kind is very economical as far as upkeep is concerned, as tools may be made up in a strip and replaced without trouble when necessary. No grinding is required except on the upper faces.

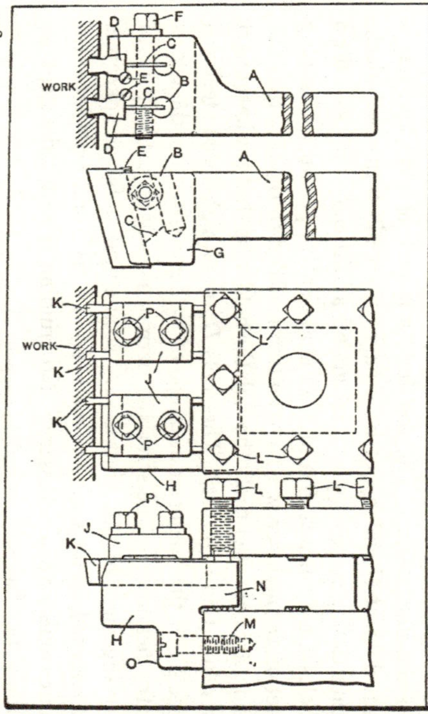

Fig. 2. Types of External Grooving Tools

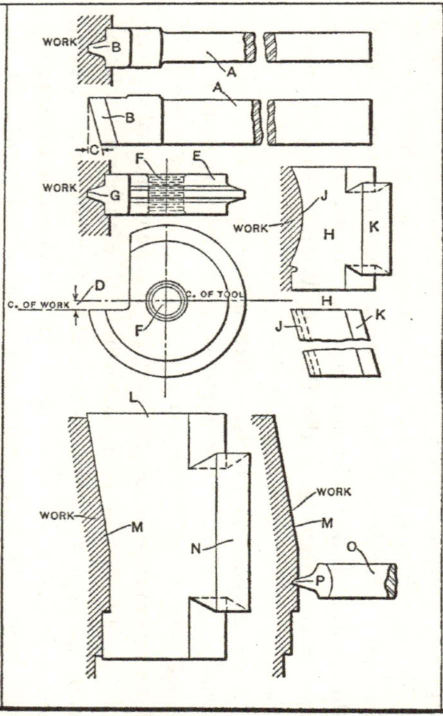

Fig. 3. Different Types of Forming Tools

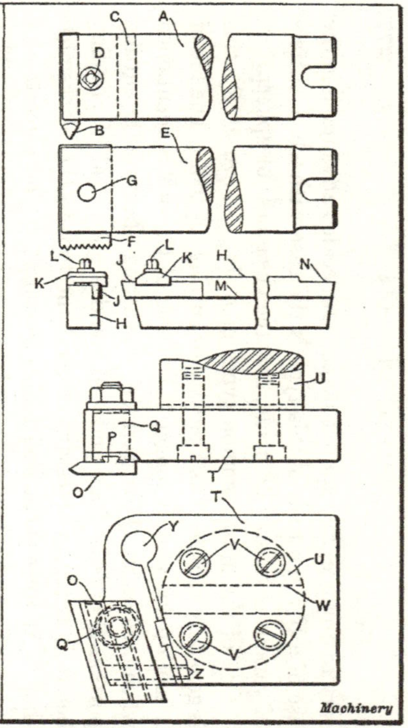

Fig. 4. Forms of Threading Tools

The lower portion of Fig. 2 shows a group of four narrow grooving tools *K* which are evenly spaced and secured in the block *H*. This block is designed for use in the side-head of a Bullard vertical turret lathe. The portion *N* takes the position normally occupied by a forged tool and is secured by the same screws *L* that are ordinarily used to hold the tool. Additional strength is obtained by the extension *O* of the block and the screws *M* which are threaded into the side-head turret. The tools *K* are of high-speed steel and are ground flat on the sides to fit the slots in the block. The steel clamp-blocks *J*, with screws *P*, are used to hold the tools in place. The under side of the clamping block is relieved to gain a good bearing on the tools. Blocks and tools of similar arrangement will be found useful for many purposes, and the cost of upkeep is low.

Forming Tools. — Tools which are so shaped as to produce forms of various kinds are more subject to changes caused by regrinding than any of the other types. Take the example shown in the upper portion of Fig. 3, in which tool *A* is formed at *B* to cut an angular groove and recess. It will be seen that this tool must be re-ground on the face, and as there must be a slight side clearance to prevent rubbing on the side of the tool, changes in the width of the grooves will take place after the tool has been ground. In addition, there will also be changes in the angle of the groove unless the tool is shimmed up to bring the cutting edge on the center. For a very few pieces a tool of this kind can be used with satisfactory results, but if much work is to be done it is better to make a circular forming tool as shown at *E*. For turret lathe work a special toolpost for either the front or rear of the cut-off slide is used in connection with this type of forming tool. A screw in the tapped hole *F* is used to draw it back and hold it against the finished surface of the toolpost. In order to give proper clearance the cutting edge is offset at *D* an amount varying from $\frac{1}{8}$ to $\frac{1}{4}$ inch, depending on the machine and the size of the work.

For work of larger size, such as that shown at *J*, the dovetail type of tool may be used. Tools of this kind are planed or shaped to the desired form and are set in a special toolpost at an angle of 15 degrees to obtain the necessary clearance. The dovetailed portion fits the tool-holder at *K* and is clamped in position by dovetail hook-bolts. It can be raised or lowered to bring the cutting edge on the center. A form such as that shown can be readily shaped and held to size without trouble if a tool of this type is used. Another dovetail tool is shown at *L*, where the form *M* is to be produced. The dovetail portion *N* fits the toolpost as in the former case. There is a groove to be cut in the work in addition to the form generated by the dovetail tool, but it is not advisable to incorporate this in the same tool, as it might be blunted before the wider faces become dull, so that the whole tool would require re-grinding on this account, which would lessen its life to a great extent. The tool *O* is therefore made to take care of the groove *P*, and is used in the rear toolpost of the cut-off slide. In all cases of forming tools which have to be used for work of a similar kind, an arrangement of this kind is desirable.

Threading Tools. — Tools used for cutting threads are of many kinds. A plain forged tool is frequently used, but its upkeep is expensive when many pieces are to be threaded. On small diameters requiring internal chased threads a forged tool is sometimes necessary on account of the size of the work, but when the work is sufficiently large to permit it, other methods are preferable. For vertical turret lathe work the bar shown at *A* in the upper part of Fig. 4 is furnished by the manufacturers for either outside or inside chasing. The end of the bar is drilled at *C* and a slot cut to provide a means of clamping tool *B* in any desired position through the screw *D*. The tool itself is dovetailed to fit a slot at the end of the bar. The cost of upkeep of a tool of this kind is very small.

The bar *E* may also be designed for use in the vertical turret lathe, but the tool *F*, in this instance, is threaded to the correct pitch for the thread to be cut. A slot in the end of the bar receives the tool, and it is secured in place by the taper pin *G*.

A very useful tool for square threads is shown at *H* and is one of the products of the Pratt & Whitney Co., Hartford, Conn. The holder is of steel and is designed for use in the engine lathe toolpost. The cutters *J* may be obtained for any pitch of thread, and are ground for side clearance. The cutters rest on the shelf *M* on the tool-holder and are clamped in place by means of strap *K* and screw *L*. An elongated hole in the strap permits adjustment for various widths of tools. Either right- or left-hand threads may be cut with the same holder by simply turning it end for end. If it becomes necessary to sharpen the tool during the progress of the cut, it may be removed from the holder and replaced without disturbing the setting. For high-grade work a roughing cutter of a size slightly smaller than the finishing tool can be used with excellent results. The cost of upkeep of a tool of this kind is very small.

A "goose-neck" holder for a lathe is frequently seen, but is seldom used for turret lathe work. The purpose of a goose-neck holder is to prevent chatter by permitting the tool to spring slightly away from the work. In Fig. 4 a special block of steel *T* is tongued at *W* to fit a slot in the end of the bar *U*, which enters a tool-holder on the turret. The block is secured in place by four screws *V*. The threading tool *O* is dovetailed at *P* to fit a corresponding slot in the block. This slot is planed at an angle to give the necessary clearance to the tool. A binder stud *Q* is dovetailed to fit, and is drawn back by the nut shown. The tool is ground across the upper face only and therefore no change in shape is caused by grinding. A hole *Y* is drilled and slot *Z* cut to allow for spring. A U-shaped spring is adjustable up and down in the slot to give more or less flexibility to the "goose-neck" portion.

Starting Tools. — When a hole of any length is to be drilled in solid stock on a turret machine, it is best to use a short drill to start the hole, afterward following with the regular drill which completes the work. This short starting or spotting drill, being somewhat stiffer than a long one, starts a hole more nearly true and saves wear and tear on the more expensive twist drill. Fig. 5 shows a starting drill at *A* held in a tool-holder *B* by the two set-screws *C*, the shank of the holder being secured in the turret hole. The angle *D* on the end of the drill should be approximately 40 degrees, instead of 31 degrees which is the regular twist-drill angle. By grinding in this way, the lips of the follow-

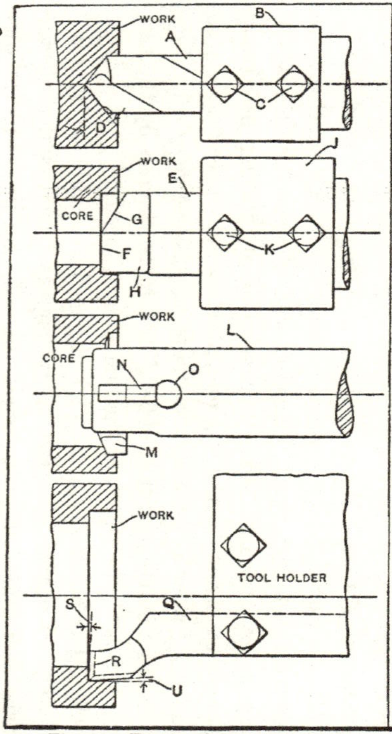

Fig. 5. Types of Starting Tools

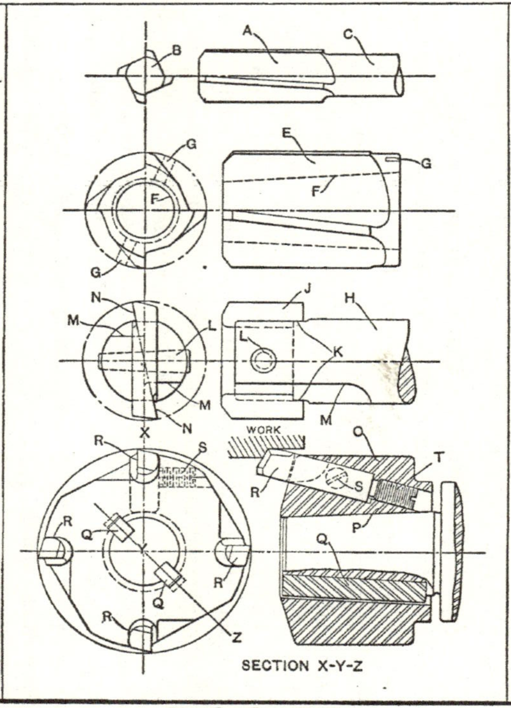

Fig. 6. Core Drills

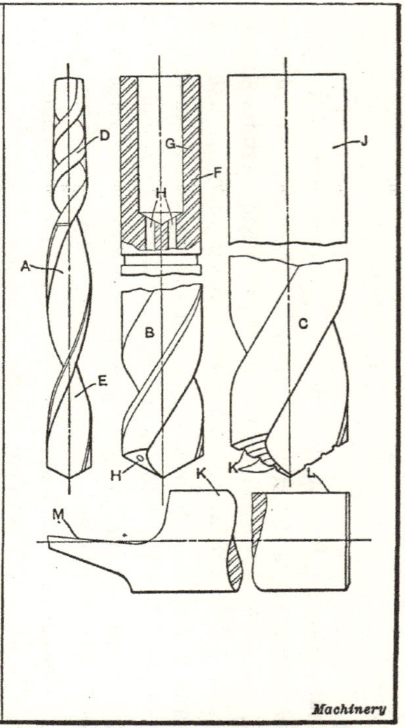

Fig. 7. Twist Drills and Drill for Brass

ing drill will start to cut before the point, thus giving them a chance to get well into the work before the point begins to guide.

In cast work, when a core drill is to be used, it is advisable to start the hole a trifle smaller than the core drill size with some sort of single-point tool, in order that the drill may start true and not be influenced by variations in the cored hole. There are several varieties of tools used for this purpose, one of these being shown at *E*. This is preferably made of high-speed steel, bright drawn if possible, and is cut down to half its diameter at *F*. It is beveled at *G* and slightly back-tapered at *H*, for clearance. It is held in a regular turret tool-holder *J* by two set-screws *K*. A tool of this kind does not deteriorate rapidly.

A chamfering tool is often used prior to a boring-bar and answers the same purpose as the tool just referred to. A combination boring- and reaming-bar is made by the Bullard Machine Tool Co., Bridgeport, Conn., for use in the vertical turret lathe. A portion of this bar is shown at *L*. The chamfering cutter *M* is of rectangular section and fits a slot in the bar at right angles to the slot *N* which is used for the floating reamer cutters. Pin *O* is flatted to act as a bearing for the latter. All cutters for this bar are of the " slip " variety, and can be readily removed or replaced by the fingers. The cost of upkeep for an arrangement of this kind is small.

When the cored hole is of fairly large size, a forged tool may be used for starting, as shown at *Q*. A tool of this type should be back-tapered at *U* to an angle of 5 degrees and should also have a slight angle at *S*, so that it will not have a tendency to ride up on the scale of the cored hole. The tool is held in a regular toolholder by set-screws. It should be so set that it will cut slightly smaller than the tool which is to follow it. It may be re-ground a number of times before re-forging becomes necessary.

Core Drills. — A type of core drill much used for small work is shown in the upper part of Fig. 6 at *A*. The shank *C* is integral with the cutting portion. The end view *B* shows one method of milling the flutes. A milling cutter should be selected which has a slightly rounded corner so that the danger of hardening cracks will be lessened. The outside of the fluted portion should be back-tapered from 0.006 to 0.010 inch per foot, in order to give the necessary clearance. Tools of this type are greatly abused in the average shop by improper grinding and by allowing them to become so dull that they cut in such a protesting manner as to be heard all over the factory. Frequently they are so re-ground that only one or two lips are cutting, thereby producing very unsatisfactory results. The shell drill shown at *E* is of the same general type, but it is made so that it fits a taper arbor at F and is driven by the slot *G* in the end, which engages with a tongue on the arbor.

Another type of tool used for the same purpose is shown at *H*. The steel shank is slotted at the end to receive the flat cutter *J*, which is shouldered at *K* to fit the flatted portion of the bar by which it is centrally located. The cutter is held in place by taper pin *L*, and is ground at *N* to bring the cutting edges on the center, or a little ahead of the center which is preferable. The bar is milled out at *M* for chip clearance. The cutter is slightly back-tapered as on a core drill. Tools of this type are much cheaper than core drills and cost less in upkeep, but they will not cut as rapidly as the four-lip type, nor will they produce work as close to size.

The cutter head shown at *O* is of steel, and was designed for rough-boring automobile cylinders. The action of this tool is much the same as that of a core drill, but replacements are much easier and cheaper. A special grinding arbor is necessary, however, so that the adjusting screws may be more accessible. The body is tapered at *P* to fit the regular arbor shown, and two keys *Q* are provided for driving. The four tools *R* are set angularly in the body, held in place by the screws *S*, and adjusted by screws *T*. The cutting lips *R* are $\frac{1}{8}$ inch ahead of the center. The body of the tool is cut away as shown, for chip clearance.

Types of Drills. — In the upkeep of drills a great deal depends upon the care used in grinding and the speeds and feeds used in their operation. The selection of the proper kind of drill to use for a particular job is also of great importance. The flat twisted drill shown at *A* in Fig. 7 is one of the products of the Pratt & Whitney Co., and is adapted for work requiring high speed and rapid production. The chip clearance on this type of drill is very generous, so that it is especially desirable for steel work. The portion *D* has an increased twist and is ground to fit a regular taper socket.

For deep-hole drilling in steel an oil drill such as that shown at *B* is useful. In this type of drill the shank *F* forms an oil pocket from which the two holes *H* lead to the drill lips. The cutting lubricant system of the machine is piped to the rear end of the drill and the lubricant is forced through the holes. This greatly assists in carrying away the heat generated and also forces the chips out along the flutes. It is sometimes necessary to speed up the pump in order to obtain sufficient pressure to force out the chips. The life of one of these drills is very long. They are one of the products of the Morse Twist Drill & Machine Co., New Bedford, Mass.

When drills are of large size, as shown at *C*, considerable power is required, and they may be made to cut to better advantage if stepped on the lips as shown at *K*. This breaks up the chips so that an easier cutting action is attained. Drills of special form, such as are used on gun barrel drilling machines, are stepped to break the chips in somewhat the same manner as that shown.

For brass or aluminum work a so-called "flat drill" is frequently used on account of the great amount of chip clearance obtained with this type. The cutting lip *M* is ahead of the center and has considerable rake. A flatted portion is milled on the shank *K* at *L*, so that the set-screws in the tool-holder will bear at this point. Tools of this kind are not very expensive and their upkeep is not costly. They should be kept very sharp and the cutting edges should be stoned smooth to secure the best results.

Cutter Heads. — Cutter heads are not usually designed unless a large number of pieces is to be machined, as they are somewhat expensive and the cost is not warranted if small lots are to be manufactured. In turret lathe practice, when inside work is to be done on a machine having turret longitudinal feed only, a cutter head may often be used to good advantage. A head designed for a condition of this kind is shown at *A* in Fig. 8. The inside of the rim is to be bored, the hub turned and the web faced.

The body of the holder is made of cast iron and is dovetailed to fit the turret face, where it is secured by means of gib *K* and screws *J*. A series of holes is drilled in the face of the head for the tools *B*, *C*, *D*, etc., these being secured in place by screws *H*. The tools are so located in the face of the head that their cutting paths overlap each other slightly so that all of the surface is

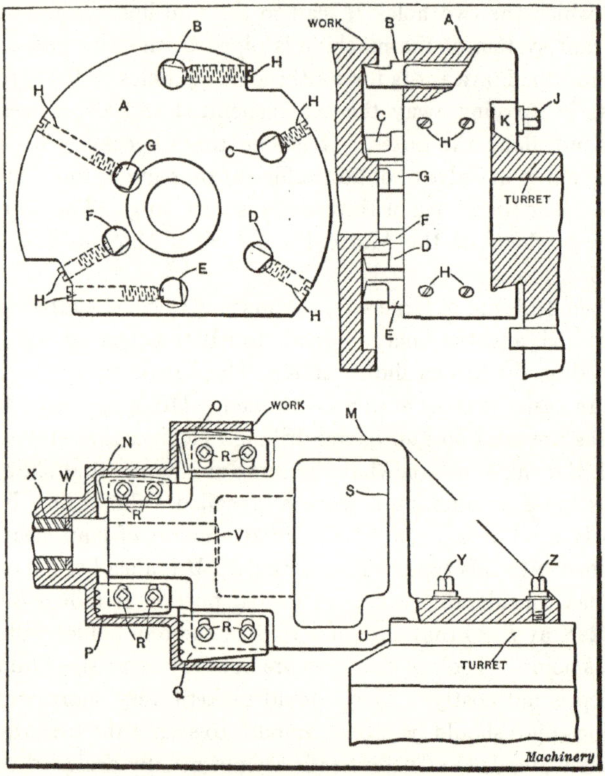

Fig. 8. Two Types of Cutter Heads

covered. Tool *B* bores the inside of the casting rim and faces a little of the web. Tool *C* faces part of the web and overlaps the cut produced by *B*. Tools *D*, *E*, F and *G* complete the remainder of the web facing and tool *G* also turns the hub. The action of this kind of tool is very satisfactory. Attention is called to the fact that the cutting faces of all the tools except *G* are ahead of the center so that they produce a scraping cut. Tool *G* has its cutting face on the center because it is used for turning and therefore should not be ahead of the center on account of its liability to " dig in."

The lower part of Fig. 8 shows another type of cutter head that was designed for boring out and facing the inside of a cone pulley. Two heads were made for this work, one for roughing and one for finishing. The heads were identical except for the facing tools, which were " nicked " in the roughing head so that they would attack the scale better. The body of the head *M* is a steel casting cored out at *S* to reduce the weight and fastened down on the turret face by screws *Y* and *Z*. Additional stiffness is secured by permitting shoulder *U* to bear against the edge of the turret. In order to prevent chatter, the forward end is provided with a revolving tool-steel bushing *X* which is a running fit on a stem forced into the body of the tool. A hardened and ground thrust washer *W* is interposed between the bushing and the shoulder on the stem. The tools themselves are of flat high-speed steel, $\frac{5}{8}$ inch thick, and are fastened in place by screws *R*. The screw holes are slotted to provide for adjustment. Tools *N* and *P*, although shown in the same plane as *O* and *Q*, are in reality set at right angles to them, so as to distribute the cut to better advantage. The facing tools *P* and *Q* are set slightly nearer the center than the boring tools *N* and *O* so that they do not cut until they reach the shoulder. Although this type of tool does not appear to be very economical as regards upkeep, it is really much better than it appears, as the tools are comparatively inexpensive. Both roughing and finishing tools are used, which is a decided advantage. There is also provision for considerable adjustment in the slotted holes.

Boring-bars. — For small work it is obvious that the cutters in boring-bars cannot be of large size, as the size of the bar will not permit it, nor can much be done in the line of adjusting screws for the same reason. About all that can be done in a case of this kind is to make up a bar something like that shown at *A* in Fig. 10. This bar has a flatted portion at *B* so that it may be firmly held by set-screws in the tool-holder. Tool *C* is of round section, flatted on one side so that the pin *D*, one side of which is flat tapered, can be driven in to hold it in place. Tools of this kind are somewhat difficult to set for diameters, but in turret lathe

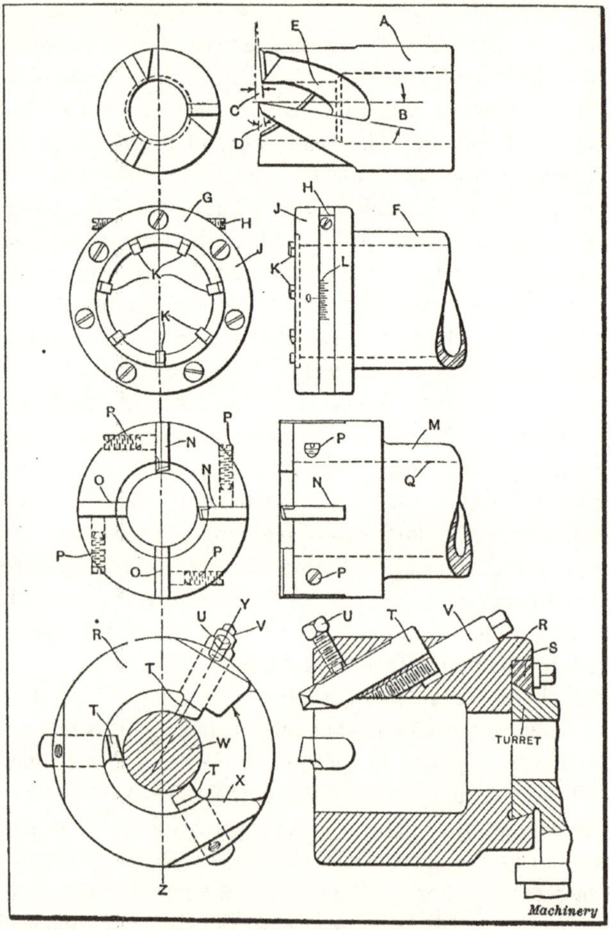

Fig. 9. Types of Hollow Mills

work, when a machine having a cross-sliding turret is used, diameters may be easily regulated by this feature.

When the work is of a size large enough to permit it, a bar such as that shown at *E* will be found much better and more convenient than that at *A*. The tool *G* is of larger section and is provided with an adjusting screw at *H*. Two set-screws *J* bear against a flat on the tool and hold it firmly. It is now possible to buy high-speed steel of round section very close to size, so that no turning is required on the tool, and the cost of upkeep of a bar of this kind is therefore very small.

For work of still larger diameter the type of bar shown at *K* will be found of advantage. This bar is slotted at the forward end, slightly larger than the thickness of some easily procurable high-speed steel of rectangular section. In order to strengthen the end of the bar a filler block is fitted to the slot and is held in place by two screws *N*. The tool is held in any desired position by the two screws *O*. No adjustment is provided for the tool. The cost of upkeep is exceptionally low.

A very different type of bar is shown at *P*, this being provided with a pilot on the forward end which enters a bushing in the chuck. The bar is made of machine steel and is provided with hardened tool steel bushings *T* which are forced onto pilot *S*. This bar is designed both for boring the large hole and facing the shoulder *Y*, as well as boring the smaller hole which extends completely through the work. The tool *R* is set at an angle of 30 degrees and is held in place by screws *U*. Screw *V* is provided for fine adjustments. The tool *Q* is set into the bar at right angles to the center line in order to facilitate grinding. Backing-up screws and set-screws are provided at *X* and *W*. This bar is cheap as regards the cost of upkeep.

Counterbores and Kindred Tools. — Fig. 11 shows at *A* a counterbore such as is regularly used for screw heads and work of a similar character. Counterbores of this type are made with three, four, five or six flutes, according to the work for which they are required, the most common being four flutes as shown at *B*. Pilots for this type of tool are frequently made removable so that special conditions can be readily met. Tools of this kind may be re-ground a number of times.

A useful tool for counterboring or cutting shoulders in larger work is shown at *E*. The body of this tool is made of machine steel and is tapered at *F* to some standard taper. Tool *H* is of

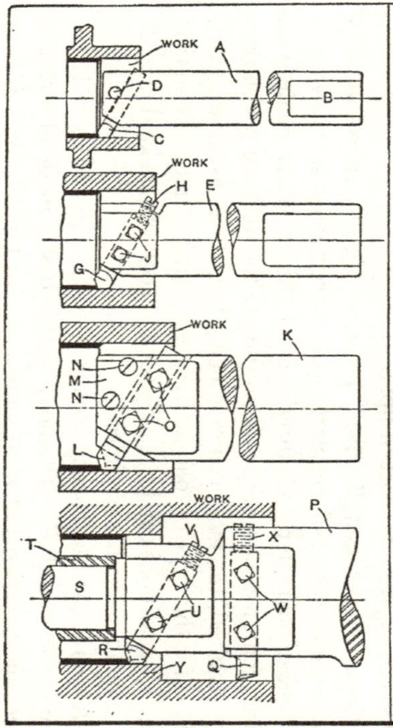

Fig. 10. Boring-bars of Different Types

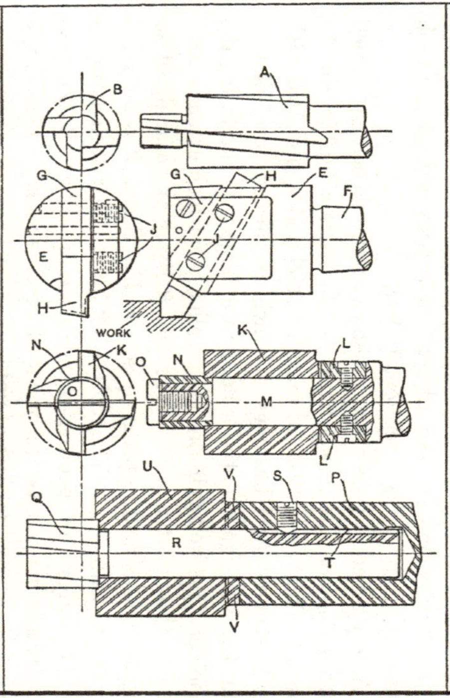

Fig. 11. Counterbores of Different Types

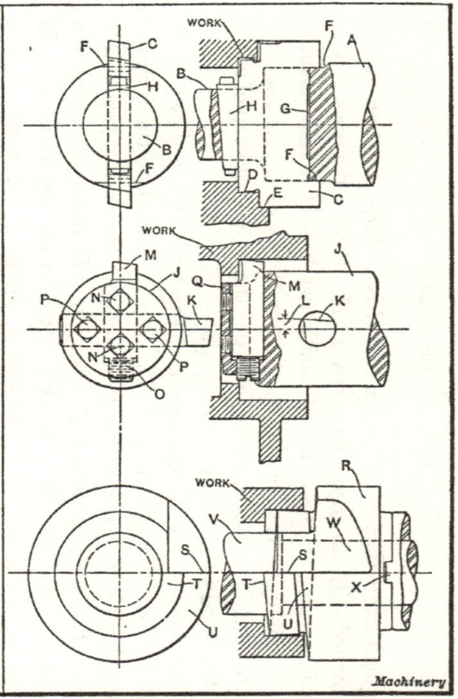

Fig. 12. Three Types of Shoulder Tools

high-speed steel and is held in position by screws *J*. The end of the holder is slotted at an angle and a filler block *G* is used to strengthen the end, being secured by a screw and dowel. Rectangular stock can be used in this holder and replacements are very easy. A holder of this kind may be used on the lathe, milling machine or turret lathe, as the occasion and conditions may demand.

A counterbore having a removable head and pilot bushing is illustrated at *K*. The stem *M* is of tool steel and is tapered at the rear end to a standard taper. Two keys are provided at *L* to drive the cutter head, these keys being let in to an enlarged portion of the stem. The cutter head *K* is shouldered at the rear end and cut out to receive the ends of the keys. The forward end of the stem is reduced in size and tapped out for screw *O* which holds the bushing *N* in place. In an arrangement of this sort a great number of combinations are possible, as heads and pilot bushings may be readily interchanged.

Another type of counterbore having interchangeable guides and heads is shown in the lower view. The construction of this tool is somewhat different from the one above. A holder *P* is slotted at *V* to act as a driver for the cutter head. The pilot *Q* is reduced in size at *R* and fits hole *T* in the shank, where it is secured by the pointed screw *S*. An angular groove prevents it from turning. The cutter head *U* is similar to that shown at *K*, and fits the stem *R* of the pilot. A tool of this type is manufactured by the Pratt & Whitney Co., Hartford, Conn., and may be bought in various sizes.

Hollow Mills. — Hollow mills are generally used on bar stock for rapid removal of metal and are usually followed by a box-tool or a single-point turner of some sort. They are not intended for accurate work. Fig. 9 shows in the upper view an ordinary type of hollow mill such as is commonly used in screw machine work. This type of tool is well known. Angle *B* varies from 10 to 15 degrees; angle *C* from 3 to 7 degrees; and angle *D* from 8 to 15 degrees. The back clearance at *E* should be approximately $\frac{1}{8}$ inch per foot, or a trifle more. The cutting edges should be on the center for brass, but a trifle ahead of the center for steel.

Mills of this type are comparatively cheap. Re-grinding changes the size somewhat, but this may be offset to a certain extent by the use of a clamping ring which draws the cutting lips together. This is not shown in the illustration.

Another form of hollow mill which possesses valuable features of adjustment is shown at *F*. This tool is made by the Geometric Tool Co., New Haven, Conn. It is not designed for heavy cutting, but is especially recommended for brass finishing and light cutting on other metals. It consists of three parts, a holder *F*, a plate *G* and a cam which controls the cutters. The holder is slotted on the front end to receive the cutters *K*; plate *G* keeps these in place. The diameters are controlled by screw *H* which operates the cutters. An index is provided on the side by which readings to 0.010 inch may be obtained. A hardened plug is recommended for setting the cutters for various diameters.

Another type of adjustable hollow mill is shown at *M*. The body of this tool is made of machine steel and is slotted to receive the two cutters *N* and the back-rests *O*. Screws *P* bind both back-rests and cutters in place when set to the required diameter. The shank *M* fits the hole in the turret. This tool is not suited for heavy cutting, but is quite adaptable to various diameters, and the cutters may be replaced with ease.

The hollow mill shown in the lower part of Fig. 9 was designed for heavy reductions on bar stock in a horizontal turret lathe. The body *R* was made of cast iron and was fitted to the turret at its rear end and drawn up by the dovetailed gib *S*. There were three tools set 120 degrees apart and placed in the holder as shown at *T*. These tools were clamped by screws *U*, while adjustment was provided by backing-up screws *V*. This tool was followed by a single-point turning tool in order to hold the size within the required limits. An opening at the side of the tool was provided at *X* to allow the egress of chips. A plug gage was made to facilitate the setting of the tools and an angular gage for the end of the tools was also found necessary in order that all tools might remove an equal amount of stock. Under test this tool reduced two-inch bar stock to one inch diameter at a speed of 100 feet per minute and a feed of 0.040 inch per revolution.

Shoulder Tools. — Great difficulty is sometimes experienced in turret lathe work in keeping shoulder distances uniform after the tools have once been ground. It is therefore advisable, whenever possible, to so design the tools that changes will not take place when they are re-ground. The desired result may sometimes be accomplished by the use of a grinding gage, but it is much better to so arrange the tool that no changes are possible. Several types of shoulder tools are shown in Fig. 12. The one at *A* is piloted with a stem *B* at the forward end in a chuck bushing. A double-ended cutter *C* of high-speed steel fits a slot in the bar and is located sideways on the two flats F. A wedge *H* holds the tool back against the rear end of the slot, and all tendency to rock is prevented by the relief at *G*. The faces of the two shoulders *D* and *E* are the important points on this piece of work, and a gage was found necessary in order to keep them uniform when grinding.

Another method of handling a somewhat similar condition is shown in the bar *J*. Two cutters of round section are inserted in the bar at right angles to each other as shown at *M* and *K*. The distance *L* is $\frac{3}{16}$ inch, so that this tool would be ahead of the center and would therefore have a scraping action in facing the end of the hub. The tool *M* used for internal boring and for facing the shoulder is also slightly ahead of the center. The screws *N* hold the tool *M*, while those at *P* secure *K*. The heads of the two screws *N* project about one-half inch more beyond the end of the bar than the other two in order to facilitate the use of a wrench. Some difficulty is experienced in keeping the shoulder distances within the required limit.

A tool which is not affected by re-grinding as far as shoulder distances are concerned is shown at *R* in the lower portion of Fig. 12. In this particular instance, a pilot *V* was used and this extended back through the holder at *W* and acted as a clamp for the cutter head, being drawn up against it by means of a nut and washer at the rear end of the holder, not shown in the illustration. The holder is provided with a driver at *X* which enters a slot at the rear of the tool proper. The tool *R* is of high-speed steel and is turned to the proper diameters and then given a slight back clearance of about 0.0015 inch to the inch to prevent rubbing. The two cutting faces *U* and *T* are faced with a flat tool on a right-hand spiral having a lead of $\frac{1}{4}$ inch to the foot. It will be seen that radial re-grinding may be easily accomplished at *S* without changing the shoulder distances. There is consequently no danger of variation.

Bottoming Tools. — In automobile construction, ball bearings are often used in the hubs, and the diameters and seats must be carefully machined. As it is not good practice to grind a reamer to a radius such as that shown in the upper portion of Fig. 13, it is necessary to provide some other tool for this work. A simple type is shown in the upper figure, this consisting of a bar *A* having a cutter of cylindrical section, formed at *B* to the proper radius. The cutter is held in place by two screws *C*, adjustment being provided by the backing-up screw *D*. The upkeep for a tool of this kind is expensive, especially if the work on which it is used is a steel casting. Sand spots and blow-holes in this class of material make frequent re-grinding necessary and considerable care must be used in re-setting in order to keep the work within the required limits of accuracy.

Another method of handling a piece of similar work is by means of a face mill having four cutting lips F formed to the radius of the seat. The mill *E* is relieved at the rear end and is driven by a flatted pin in the holder *G*. The pilot *H* enters a bushing in the chuck and is shouldered at *J* to assist in keeping the mill in position. The shank passes entirely through the holder and is secured at the rear end by a nut and washer. A tool of this type may be re-ground considerably without seriously changing the contour. A point worthy of note in the design of face mills which are hand-operated (as in turret lathe practice) is that the lips or cutting faces should be as few as is consistent with the work in question, especially if much surface is to be faced. Too many lips make it difficult to make the mill cut. The mill shown in the illustration was originally designed with eight cutting lips, but it was found that it was about all a man could do to force it to cut on account of the amount of surface to which the cutting edges were presented at once. Four of the lips were therefore ground away; this overcame the difficulty. The approximate diameter of this mill was $3\frac{1}{2}$ inches; it was used on a horizontal turret lathe.

An arrangement for a bottoming tool that can be re-ground without changing its shape is shown in the lower illustration. A bar *K* is held in the turret and is beveled on the forward end to an angle of 30 degrees, and is counterbored and tapped off center, as shown at *M* and *N*, to receive a shouldered screw which holds the circular forming tool *L* in position. Attention is called to the fact that the center of this tool is $\frac{1}{8}$ inch above the center of the work in order to provide the necessary clearance. It will be noted that the form *O* does not change its shape, because the tool is ground radially. The expense of making a tool of this kind is not excessive, nor is the cost of upkeep.

Straight Chucking Reamers. — There are many varieties of adjustable reamers on the market, but very few can be adjusted to cut a certain size without the necessity of re-grinding. Re-grinding, however, is a small matter when compared with the cost of a new reamer. Probably the simplest type of reamer that will produce satisfactory work is that shown in the upper illustration of Fig. 14. The bar *A* fits the turret hole and is slotted at the forward end to receive the flat reamer cutter *B*. The back of this tool is relieved at *F* and bears against pin *E* which is of hardened steel, flattened on one side, and forced into the bar. The screw *C* has a teat *D* which fits loosely into a hole in the reamer cutter, so that a certain amount of "float" is permitted. A tapered cut in tool *B* runs into the hole which the teat screw enters, and a plug *R* may be forced down to open up the blade, thus permitting a number of re-grindings. A reamer of this sort, when used on a horizontal turret lathe, should be turned so that the blade stands vertical in order to minimize errors in the indexing of the turret. It should be lightly guided by the fingers in entering the hole.

The type of reamer shown at *G* is non-adjustable, except if re-ground. In this type of reamer shank *G* is held at the rear end in a floating holder. The forward end is milled out so that a number of blades as shown at *H* may be inserted in the slots and

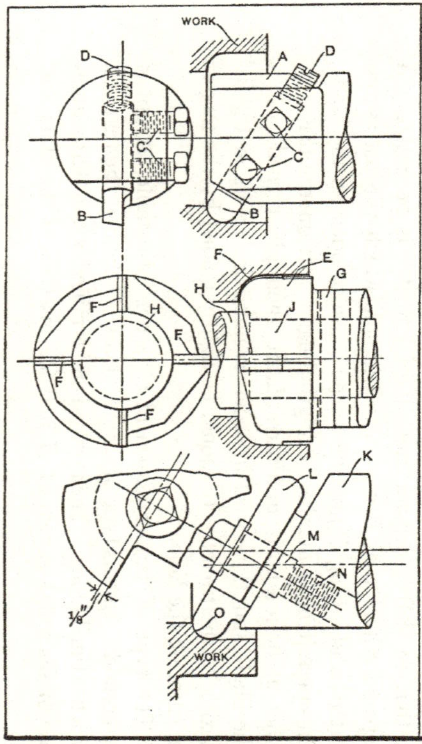

Fig. 13. Bottoming Tools

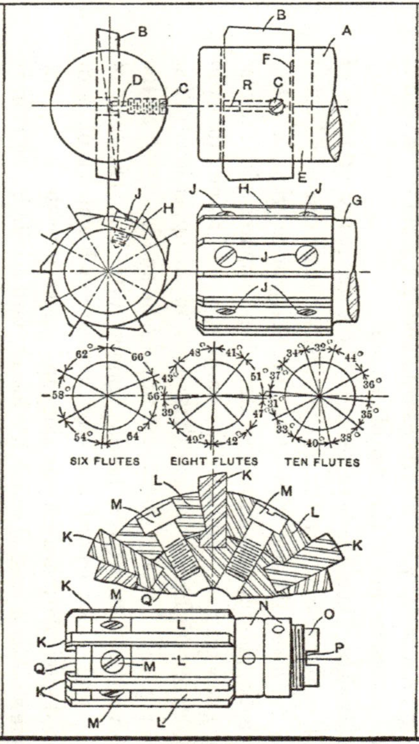

Fig. 14. Finishing Reamers

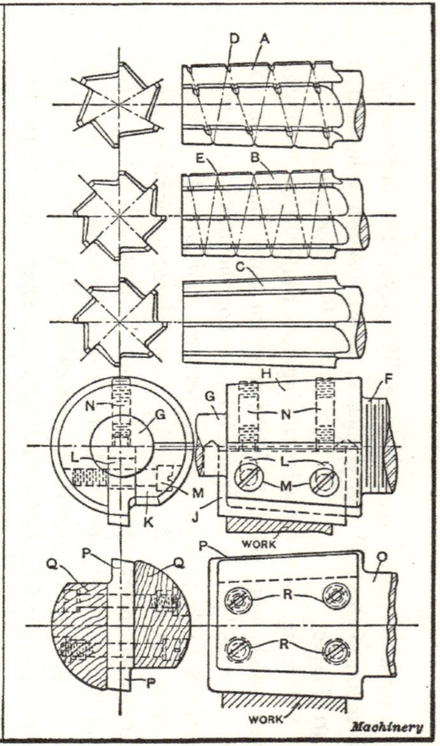

Fig. 15 Taper Reamers

fastened by screws *J*. In milling a reamer of this kind, the blades are inserted and the cutting done on body and blade together. The blades are then taken out and hardened, after which they are assembled and ground in position, both cylindrically and along the faces of the flutes. When a reamer of this kind becomes worn, the blades may be removed and "papered up," and then re-ground. The cost of upkeep on this type of reamer is small.

Occasionally it is desirable to space flutes unequally in order to obviate chatter. It is contended by some that unequal spacing on one-half of the reamer answers every purpose and that it only causes extra trouble to space every flute differently; in other words, in an even-fluted reamer the teeth may be opposite each other, thus facilitating calipering; this is, however, doubtful. Unequal spacing for six-, eight- and ten-flute reamers is given in the diagram, and other spacings may be readily determined.

The expansion type of reamer shown in the lower portion of Fig. 14 is made by the Pratt & Whitney Co., Hartford, Conn. A body *Q* fits a taper arbor and is driven by the slot *P* which engages with a pin in the arbor. The blades *K* fit the tapered slots shown in the sectional view. The clamps *L* are drawn down to lock the blades securely by means of the screws *M*. The blades may be set for various diameters within their capacity by means of the two check-nuts *N*. This type of reamer is exceptionally good for turret lathe work.

Taper Reamers. — Taper reamers should always be made considerably longer than the hole which they are to ream in order to provide for long life. In rough-cored taper holes a starting tool or boring-bar should first be used in order to start a true hole, so that the roughing taper reamer may not be influenced by the irregularities of the core. The series of reamers *A*, *B* and *C* shown in Fig. 15 are used for roughing, rough-finishing and final finishing; *A* has six flutes and a series of notches as shown at *D* in order to break the chips and make cutting easier; reamer *B* has eight flutes and also a right-hand thread or groove *E*; the finishing reamer *C* also has eight flutes and may or may not have a groove like that at *E*. If the lead of this groove is not sufficiently great, a series of steps will appear in the finished work. If the taper is very shallow, it may be found necessary to mill the flutes left-hand in order to prevent " drawing-in " and chatter.

A taper scraping tool having a single blade capable of considerable adjustment is shown at *H*. A tool of this type is preferably piloted at *G* in a bushing in the chuck while the rear end is secured firmly in the turret hole. A slot is cut in the body in which the blade *J* is clamped by means of the two screws *M*. These screws pass through the slotted holes *L*, thereby permitting radial adjustment. Two backing-up adjusting screws *N* bear against the back of the blade. The body is cut away at *K* to provide for chip clearance. Shank F may or may not be threaded for a collar so as to provide a longitudinal stop. This reamer should be held in such a way that the blade will stand vertically when used in horizontal turret lathe work. This tool has given excellent results.

A cheap taper reamer with which good work can be done is shown at *P* in the lower portion of Fig. 15. The blade *P* is flat and is formed to the desired taper. The portion *O* is permitted to float in a slotted holder driven by a pin. The two pieces *Q* are of maple or some other wood equally good, and are formed to the correct taper, so that they follow the blade and prevent chatter. They are fastened in place by the four screws *R* which pass completely through the blade and enter bushings on the opposite side.

INDEX

Spotted something that isn't quite right?

Let us know and we'll fix it for future reprints. And as a thank you for your diligence, we'll send you a free Sothis book of your choice.

publisher@sothis.press

This page intentionally left blank

Made in United States
Orlando, FL
05 November 2026

90229007R00187